MW01071284

REAL ANALYSIS AND APPLICATIONS

Including Fourier Series and the Calculus of Variations

REAL ANALYSIS AND APPLICATIONS

INCLUDING FOURIER SERIES AND THE CALCULUS OF VARIATIONS

FRANK MORGAN

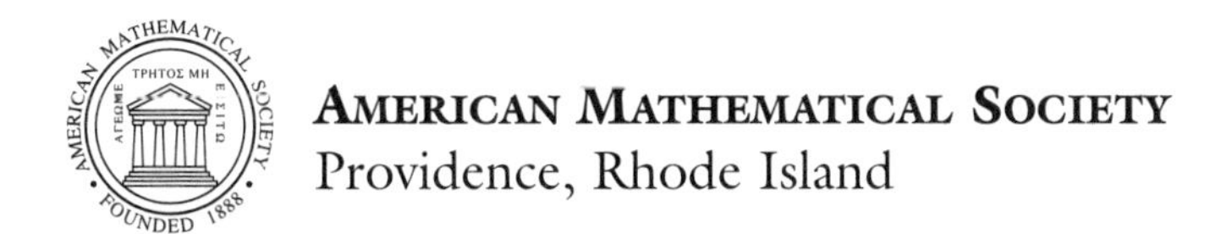
AMERICAN MATHEMATICAL SOCIETY
Providence, Rhode Island

2000 *Mathematics Subject Classification.* Primary 26–01, 49–01, 42–01, 83Cxx.

Front cover: The cover illustrates how continuous functions can converge nonuniformly to a discontinuous function. It is based on Figure 16.1, page 76. The rotating ellipses in the background suggest the long mysterious precession of Mercury's orbit, finally explained by Einstein's General Relativity (Chapter 37), a nice application of real analysis.

Back cover: The author, in the Math Library outside his office at Williams College. Photo by Cesar Silva.

Cover design by Erin Murphy of the American Mathematical Society, on a suggestion by Ed Burger.

For additional information and updates on this book, visit
www.ams.org/bookpages/realapp

Library of Congress Cataloging-in-Publication Data

Morgan, Frank.
Real analysis and applications : including Fourier series and the calculus of variations / Frank Morgan.
p. cm.
Includes bibliographical references and index.
ISBN 0-8218-3841-5 (alk. paper)
1. Mathematical analysis—Textbooks. 2. Calculus of variations—Textbooks. 3. Functions of real variables—Textbooks. 4. Fourier series—Textbooks. I. Title.

QA315.M58 2005
515—dc22 2005054563

∞ The paper used in this book is acid-free and falls within the guidelines established to ensure permanence and durability.
Visit the AMS home page at `http://www.ams.org/`

10 9 8 7 6 5 4 3 2 1 10 09 08 07 06 05

Contents

Preface

Our lives and the universe barely work, but that's OK; it's amazing and great that they work at all. I think it has something to do with math, and especially real analysis, the theory behind calculus, which just barely works. Did you know that there are functions that are not the integral of their derivatives, and that a function with a positive derivative can decrease? But if you're a little careful you can get calculus to work. You'll see.

The theory is hard, subtle. After Newton and Leibniz invented the calculus in the late 1600s, it took puzzled mathematicians two hundred years, until the latter 1800s, to get the theory straight. The powerful modern approach using open and closed sets came only in the 1900s. Like many others, I found real analysis the hardest of the math major requirements; it took me half the semester to catch on. So don't worry: just keep at it, be patient, and have fun.

The applications in the calculus of variations are amazing, from computing optimal economic strategies to predicting the relativistic correction to Mercury's orbit.

This text is designed for students. It presents the theoretical intellectual breakthroughs which made calculus rigorous, but always with the student in mind. If a shortcut or some more advanced comments without proof provide better illumination, we take the shortcut and make the comments. The result is a complete course on real analysis that fits comfortably in one semester. Chapters 1–17 provide the theoretical core, which can for example be supplemented by Chapters 18–20 and 23–25 (Fourier Series) or by selections from Part V on the Calculus of Variations.

This text developed with a one-semester undergraduate analysis course at Williams College. I would like to thank my colleagues and students, especially Ed Burger, Tom Garrity, Kris Tapp, Nasser Al-Sabah '05, and Matt Spencer '05, and my editors Ed Dunne and Tom Costa. Other texts I found helpful for applications include *Dynamic Optimization* by Kamien and Schwartz and *Methods of Applied Mathematics* by Hildebrand; see also my own *Riemannian Geometry.*

Frank Morgan
Department of Mathematics and Statistics
Williams College
Williamstown, Massachusetts
www.williams.edu/Mathematics/fmorgan
Frank.Morgan@williams.edu

Part I

Real Numbers and Limits

Chapter 1

Numbers and Logic

1.1. Numbers. Calculus and real analysis begin with numbers:

The *natural numbers*

$$\mathbb{N} = \{1,\ 2,\ 3, \ldots\}.$$

The *integers*

$$\mathbb{Z} = \{\ldots,\ -3,\ -2,\ -1,\ 0,\ 1,\ 2,\ 3,\ \ldots\}$$

($\mathbb{Z}$ stands for the German word Zahl for number).

The *rationals*

$$\begin{aligned} \mathbb{Q} &= \{p/q \text{ in lowest terms: } p \in \mathbb{Z},\ q \in \mathbb{N}\} \\ &= \{\text{repeating or terminating decimals}\}. \end{aligned}$$

($\mathbb{Q}$ stands for quotients).

The *reals*

$$\mathbb{R} = \{\text{all decimals}\}$$

with the understanding that $.999\cdots = 1$, etc. Reals which are not rational are called *irrational.* Thus the set of irrationals is the *complement* of the set of rationals, and we write

$$\{\textit{irrationals}\} = \mathbb{Q}^{\mathsf{C}} = \mathbb{R} - \mathbb{Q}.$$

1.2. Intervals in $\mathbb{R}$. For $a < b$, define *intervals*

$$\begin{aligned} [a, b] &= \{x \in \mathbb{R} \colon a \le x \le b\}, & a \bullet\!\!-\!\!\!-\!\!\!-\!\!\bullet\, b \\ (a, b) &= \{x \in \mathbb{R} \colon a < x < b\}, & a \circ\!\!-\!\!\!-\!\!\!-\!\!\circ\, b \\ [a, \infty) &= \{x \in \mathbb{R} \colon a \le x\}, & a \bullet\!\!\longrightarrow \end{aligned}$$

and so on.

1.3. $\mathbb{R}^n$. We will consider the plane $\mathbb{R}^2$ of pairs (x, y) of real numbers and more generally n-dimensional space $\mathbb{R}^n$ of n-tuples $(x_1, x_2, \ldots, x_n)$ of real numbers. In these spaces, by rationals we'll mean n-tuples of rationals.

There are fancier number systems, such as the complex numbers (which will make a solo appearance in Chapter 21) and the much fancier "quaternions" (which maybe you'll see in some future course).

1.4. Distance. The (Euclidean) *distance* between real numbers x and y is given by $|y - x|$. More generally, the distance between two vectors (x_i), (y_i) in $\mathbb{R}^n$ is given by

$$|y - x| = ((y_1 - x_1)^2 + (y_2 - x_2)^2 + \cdots + (y_n - x_n)^2)^{1/2}.$$

Distance satisfies the *triangle inequality*, which says that the length of one side of a triangle with vertices x, y, z is less than or equal to the sum of the lengths of the other two sides:

$$|x - z| \leq |x - y| + |y - z|.$$

The Ancient Greeks discovered with dismay that not every real is rational, with the following example.

1.5. Proposition. $\sqrt{2}$ *is irrational.*

Proof. Suppose that $\sqrt{2}$ equals p/q, in lowest terms, so that p and q have no common factors. Since $2 = p^2/q^2$,

$$2q^2 = p^2.$$

Thus p must be even: $p = 2p'$. Hence

$$2q^2 = 4p'^2,$$
$$q^2 = 2p'^2,$$

which means that q must be even, a contradiction of lowest terms, since p and q both have 2 as a factor. □

There is an easier example:

1.6. Proposition. $\log_{10} 5$ *is irrational.*

Proof. Suppose that $\log_{10} 5$ equals p/q, which means that $10^{p/q} = 5$ or

$$10^p = 5^q.$$

But 10^p ends in a 0, while 5^q ends in a 5, contradiction. □

Probably the easiest way to give an example of an irrational is just to write down a nonrepeating, nonterminating decimal like

$$.01\ 001\ 0001\ 00001\ 000001\ldots.$$

although that's no number you've ever heard of.

1.7. Logic and ambiguous English. English is a treacherous language, and we have to be very careful. For example, there are two meanings of "or" in English:

(1) A crazy "exclusive" or, meaning "one or the other, but not both." I had lunch in the French Quarter, and they offered me a choice of black-eyed peas *or* greens, but they wouldn't give me both.

(2) The sensible, mathematical "inclusive" or, meaning "one or the other or both." You can work on your real analysis homework day *or* night.

1.8. Implication. There are many ways to say that one statement A implies another statement B. The following all mean exactly the same thing:

- If A, then B.
- A implies B, written $A \Rightarrow B$.
- A only if B.
- B if A, written $B \Leftarrow A$.

Here are some examples of such equivalent statements:

- If a number n is divisible by four, then it is even.
- n divisible by four implies n even.
- n is divisible by four only if it is even (never if it is odd).
- n is even if divisible by four.

Such an implication is true if B is true or if A is false (in which case we say that the implication is "vacuously true"). For example, "If 5 is even, then 15 is prime" is vacuously true. This is pure logic and has nothing to do with causation. (Indeed, from the point of view of causation, if 5 were even, then 15, as a multiple of 5, would be even and hence *not* prime.)

Such an implication is false only if A is true and B is false.

Such an implication is logically equivalent to its contrapositive: not B implies not A. This is the basis for proof by contradiction. We assume the negation of the B we are trying to prove, and arrive at a contradiction of the hypothesis A.

Such an implication is logically distinct from its converse: B implies A.

- If x is a Williams student, then x is a human being. True.
- If x is not a human being, then x is not a Williams student. Contrapositive, true.
- If x is a human being, then x is a Williams student. Converse, false in this case.

If it happens that the implication and its converse are both true, i.e., if $A \Rightarrow B$ and $B \Rightarrow A$, then we say that A and B are logically equivalent or that

$$A \text{ if and only if } B \qquad (A \Leftrightarrow B).$$

Example. Let $S \subset \mathbb{Z}$. Consider the statements

A: All elements of S are even. (For all $x \in S$, x is even.)

B: Some element of S is even. (There exists $x \in S$, such that x is even.)

Does $B \Rightarrow A$? Certainly not in general.

Does $A \Rightarrow B$? Not in general because if S is the empty set $\varnothing$, then A is true (vacuously), while B is false!

1.9. Abbreviations. Understand but avoid abbreviations such as

$$\wedge \text{ and} \quad \vee \text{ or} \quad \sim \text{ not} \quad \exists \text{ exists} \quad \forall \text{ for all.}$$

1.10. Sets. A set A is completely determined by the elements it contains:

$$A = \{x \colon x \in A\}.$$

If all elements of A are elements of B, we say that A is a *subset* of B ($A \subset B$) or that B *contains* A ($B \supset A$). You have to be careful, because the word "contains" is used both for subsets and for elements, as determined by context. Proper containment ($A \subset B$ and $A \neq B$) is denoted by $A \subsetneq B$. Some texts denote ordinary containment by $A \subseteq B$ to remind you that equality is allowed. We say that sets A, B *intersect* if $A \cap B \neq \varnothing$. The symbol $\bigcap_{n=1}^{\infty} A_n$ denotes the set of elements common to all of the sets $A_1, A_2, A_3, \ldots$:

$$x \in \bigcap_{n=1}^{\infty} A_n \quad \text{if and only if} \quad x \in A_n \text{ for all } n.$$

We define the complement of a set

$$A^{\mathsf{C}} = \{x \colon x \notin A\},$$

and the difference of two sets

$$A - B = \{x \in A \colon x \notin B\}.$$

1.11. Functions. A *function* $f\colon A \to B$ takes an input x from a domain A and produces an output $f(x)$ in some range B. For most of this book we'll assume that $A \subset \mathbb{R}^n$ and $B \subset \mathbb{R}$. The set of all outputs

$$f(A) = \{f(x)\colon x \in A\} \subset B$$

is called the *image* of f. If the image is the whole range, then f is called *onto* or *surjective*. If f maps distinct points to distinct values, then f is called *injective* or *one-to-one* (1-1). If f is both injective and surjective, then f is called *bijective* or a *1-1 correspondence* and f has an inverse function $f^{-1}\colon B \to A$.

If X is a subset of A, then $f(X)$ is the corresponding set of values:

$$f(X) = \{f(x)\colon x \in X\}.$$

Whether or not f^{-1} exists as a function on points, for a set $Y \subset B$, you can always take the *inverse image*

$$f^{-1}Y = \{x\colon f(x) \in Y\}.$$

For example, if $f(x) = \sin x$, then

$$f^{-1}\{0\} = \{0, \pm\pi, \pm 2\pi, \pm 3\pi, \dots\}.$$

Exercises 1

1. Is the statement

$$\text{If } x \in \mathbb{Q}, \text{ then } x^2 \in \mathbb{N}$$

true or false for the following values of x? Why?

a. $x = 1/2$.

b. $x = 2$.

c. $x = \sqrt{2}$.

d. $x = \sqrt[4]{2}$.

2. For which real values of x is the converse of the statement of Exercise 1 true?

3. Which of the following statements are true of the real numbers.

a. For all x there exists a y such that $y > x^2$.

b. There exists a y such that for all x, $y > x^2$.

c. There exists a y such that for all x, $y < x^2$.

d. $\forall a, b, c$, $\exists x$ such that $ax^2 + bx + c = 0$.

4. Which of the following statements are true for all real numbers x.

a. If $x \in (1,2]$, then $x^2 \in (1,4]$.

b. If $x \in (-1,2]$, then $x^2 \in (-1,4]$.

c. If $x \in (-1,2]$, then $x^2 \in (1,4]$.

5. a. What is the length of a diagonal of a unit square (with vertices $(0,0)$, $(0,1)$, $(1,0)$, $(1,1)$)?

b. What is the length of a long diagonal of a unit cube?

c. What is the length of the longest diagonal of a unit hypercube in $\mathbb{R}^4$?

d. What is the length of the longest diagonal of a unit hypercube in $\mathbb{R}^n$?

6. Prove that $\sqrt[4]{2}$ is irrational.

7. Prove that $\log_{10} 15$ is irrational.

8. Find infinitely many nonempty sets of natural numbers

$$\mathbb{N} \supset S_1 \supset S_2 \supset S_3 \supset \cdots$$

such that $\bigcap_{n=1}^{\infty} S_n = \varnothing$.

9. Identify each of the following functions f from $\mathbb{R}$ to $\mathbb{R}$ as injective (1-1), surjective (onto), neither, or both (bijective).

a. $f(x) = -x$.

b. $f(x) = x^2$.

c. $f(x) = \sin x$.

d. $f(x) = e^x$.

e. $f(x) = x^3 + x^2$.

10. Give a counterexample to one of the following four formulas for images and inverse images of sets (the other three are true):

$$f(X_1 \cup X_2) = f(X_1) \cup f(X_2), \quad f^{-1}(Y_1 \cup Y_2) = f^{-1}(Y_1) \cup f^{-1}(Y_2),$$
$$f(X_1 \cap X_2) = f(X_1) \cap f(X_2), \quad f^{-1}(Y_1 \cap Y_2) = f^{-1}(Y_1) \cap f^{-1}(Y_2).$$

Chapter 2

Infinity

Although some infinite sets contain other infinite sets, there is a good intuition that in some sense all infinite sets are the same size. For example, although the even natural numbers are a proper subset of all the natural numbers, the two sets can be paired up by matching $2n$ with n:

$$\begin{array}{cccccc} 2, & 4, & 6, & 8, & 10, & \dots \\ 1, & 2, & 3, & 4, & 5, & \dots \end{array}$$

Although many infinite sets are the same size as the natural numbers, it turns out that some really are much bigger in a mathematically precise sense.

2.1. Definition. A set is called *countable* if it is finite or it can be put in 1-1 correspondence with the natural numbers, i.e., if its elements can be listed.

Examples of countable sets include the set $3\mathbb{N}$ of positive multiples of 3 and the set $\mathbb{Z}$ of all integers (including the negative ones):

$\mathbb{N}$	$3\mathbb{N}$	$\mathbb{Z}$
1	3	0
2	6	1
3	9	-1
4	12	2
5	15	-2

It is clear that *a subset of a countable set is countable* because you can use the order of the list of the set to list the subset. It turns out that even the apparently larger set of all rationals is still countable.

2.2. Proposition. *The set* $\mathbb{Q}$ *of rationals is countable.*

Proof. It suffices to show that the positive rationals are countable; then the lists of positive and negative rationals and zero can be interspersed on a single list as for $\mathbb{Z}$ above. Arrange all ratios in an infinite table like this:

$$\begin{array}{cccccc} 1/1 & 1/2 & 1/3 & 1/4 & 1/5 & \dots \\ 2/1 & 2/2 & 2/3 & 2/4 & 2/5 & \dots \\ 3/1 & 3/2 & 3/3 & 3/4 & 3/5 & \dots \\ 4/1 & 4/2 & 4/3 & 4/4 & 4/5 & \dots \\ 5/1 & 5/2 & 5/3 & 5/4 & 5/5 & \dots \\ \dots & \dots & \dots & \dots & \dots & \dots \end{array}$$

Now starting at the upper lefthand corner, move through the northeasterly diagonals of this table, starting with 1/1, then 2/1 and 1/2, then 3/1, 2/2, and 1/3, and so on.

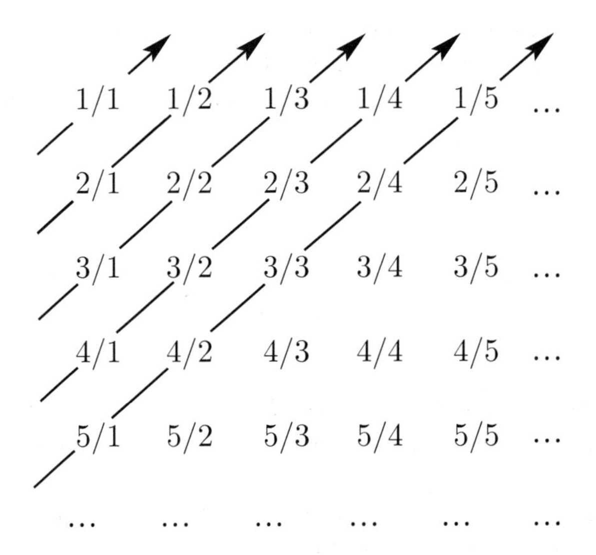

When you hit a repetition (like $2/2 = 1/1$), skip it. This process puts the rationals on a list:

$$1/1,\ 2/1,\ 1/2,\ 3/1,\ 1/3,\ 4/1,\ 3/2,\ 2/3,\ 1/4,\ 5/1,\ 1/5,\ \dots.$$

Therefore, the rationals are countable. □

2.3. Proposition. *The Cartesian product of two countable sets*

$$C_1 \times C_2 = \{(c_1, c_2) \colon c_1 \in C_1 \text{ and } c_2 \in C_2\}$$

is countable. A countable union of countable sets is countable.

(A "countable union" means "the union of countably many.")

Proof. The proofs (Exercises 4 and 5) are very similar to the proof of Proposition 2.2. □

By now you may be thinking that all sets are countable. Georg Cantor showed in the late 1800s that the reals $\mathbb{R}$ are not countable. This sounds very hard to prove. How could you show that there is no 1–1 correspondence between $\mathbb{R}$ and $\mathbb{N}$? The proof is amazing.

2.4. Proposition. *The set $\mathbb{R}$ of reals is uncountable.*

Proof. We will assume that Proposition 2.4 is false and derive a contradiction. In particular we will assume that the reals can be listed, and then exhibit a real number missing from the list, the desired contradiction. Since the argument applies to any list, that will complete the proof.

Suppose that the reals were countable. Then the positive reals would be countable and could be listed, for example:

$$\begin{array}{lr} 1. & 657.\mathbf{8}53260\ldots \\ 2. & 2.3\mathbf{1}3333\ldots \\ 3. & 3.14\mathbf{1}592\ldots \\ 4. & .000\mathbf{2}07\ldots \\ 5. & 49.4949\mathbf{4}9\ldots \\ 6. & .87325\mathbf{7}\ldots \\ \ldots & \end{array}$$

To obtain a contradiction, we just have to show that some real α (Greek letter alpha, see Table of Greek letters on page 193) has to be missing from this list. We'll construct such an α by making its first decimal place different from the first decimal place of the first number on the list, by making its second decimal place different from the second decimal place of the second number on the list, and in general by making its nth decimal place different from the nth decimal place of the nth number on the list. To be specific, we'll make the nth decimal place 1, unless the nth decimal place of the nth number on the list is 1, in which case we'll make it 2. For our example,

$$\alpha = .122111\ldots$$

As promised, we have found a number α missing from the list, because it differs from the nth number on the list in the nth decimal place. Since this argument applies to any list, the reals cannot be listed. $\square$

2.5. Uncountable infinities. More generally, one can say that two infinite sets have the same size if they can be put in 1-1 correspondence. It turns out that there is an infinite hierarchy of infinitely many different size infinities. What's bigger than $\mathbb{R}$? Not $\mathbb{R}^2$ or $\mathbb{R}^n$; they turn out to be the same infinity as $\mathbb{R}$. One bigger infinity is the set of all functions from $\mathbb{R}$ to $\mathbb{R}$. Another is the set of all subsets of $\mathbb{R}$. Indeed, the set of all subsets of any set X is always bigger than X, so this is one way to keep going. The proof, which you could find on the internet, is short but tricky.

Exercises 2

1. Prove whether or not the set S is countable.

a. $S = \{\text{irrationals}\}$.

b. $S = \{\text{terminating decimals}\}$.

c. $S = [0, .001)$.

d. $S = \mathbb{Q} \times \mathbb{Q}$.

e. $S = \mathbb{R} \times \mathbb{Z}$.

2. Prove that the intersection of two countable sets is countable.

Hint: There is a very short answer.

3. Prove whether or not the intersection and union of two uncountable sets must be uncountable.

4. Prove that the Cartesian product of two countable sets

$$A \times B = \{(a, b) \colon a \in A \text{ and } b \in B\}$$

is countable.

Hint: Imitate the proof of Proposition 2.2.

5. Prove that a countable union of countable sets is countable.

Hint: Imitate the proof of Proposition 2.2.

Chapter 3

Sequences

Many would say that the hardest theoretical concept in analysis is *limit*. What does it mean for a sequence of numbers to converge to some limit? There just is no easy answer. Oh, we'll try to find one, but there will always be some sequences that a simple answer cannot handle. In the end, we'll be forced to do something a little complicated, and to make it worse, we'll follow tradition and use the Greek letter ε (epsilon).

3.1. Discussion. The sequence

$$1,\ 1/2,\ 1/3,\ 1/4,\ 1/5,\ \ldots \tag{1}$$

converges to 0. The sequence of digits of π

$$3,\ 1,\ 4,\ 1,\ 5,\ 9,\ 2,\ \ldots \tag{2}$$

does not converge to anything; it just bounces around. The sequence

$$1,\ 2,\ 3,\ 4,\ 5,\ \ldots \tag{3}$$

diverges to infinity. Those sequences are easy. But sometimes it is hard to decide. What about the sequence

$$1,\ 0,\ 1,\ 0,\ 0,\ 1,\ 0,\ 0,\ 0,\ 1,\ 0,\ 0,\ 0,\ 0,\ \ldots ? \tag{4}$$

Common agreement is that it does not converge, but to decide we really need a good definition of "converge." How about this:

First attempt at a definition. A sequence

$$a_1,\ a_2,\ a_3,\ a_4,\ a_5,\ \ldots$$

converges to, say, 0 if the terms get closer and closer to 0.

According to this definition, (4) does not converge. Good, but what about the following sequences:

(5) $$8,\ 1,\ 4,\ 1/2,\ 2,\ 1/4,\ 1,\ 1/8,\ 1/2,\ 1/16,\ 1/4,\ 1/32,\ \ldots$$

(6) $$1\frac{1}{2},\ 1\frac{1}{4},\ 1\frac{1}{8},\ 1\frac{1}{16},\ 1\frac{1}{32},\ \ldots$$

The terms of (5) are not getting "closer and closer to 0," but the sequence does converge to 0. The terms of (6) are getting "closer and closer to 0," but the sequence does not converge to 0. (We will see that this sequence converges to 1.)

We need a more precise definition. The terms need to get *and stay* arbitrarily close to zero or whatever limit value L, *eventually.* "Arbitrarily close" means as close as anyone could prescribe, i.e., given any positive error, the terms eventually have to stay within that tolerance of error. Since the letter e is already taken by 2.718281828..., mathematicians usually use the Greek letter epsilon ε for the given error. And what do we mean by "eventually"? We mean that given the tolerance of error ε, we can come up with a big number N, such that all the terms after a_N are within ε of the limit value L. That is, given $\varepsilon > 0$, we can come up with an N, such that whenever $n > N$, every subsequent a_n is within ε of L. Now that's a good definition, and here it is written out concisely:

3.2. Definition of convergence. A sequence a_n *converges* to a *limit* L

$$a_n \to L$$

if given $\varepsilon > 0$, there is some N, such that whenever $n > N$,

$$|a_n - L| < \varepsilon.$$

Otherwise we say that the sequence *diverges.*

Notice that order in the definition is very important. First comes the sequence a_n and the proposed limit L. Second comes the tolerance of error ε, which is allowed to depend on the sequence. Third comes the N, which is allowed to depend on ε.

A sequence which diverges might diverge to infinity like (3) or diverge by oscillation like (2).

Some other things in the definition do not matter, such as whether the inequalities are strict or not. For example, if you knew only that you could get $|a_n - L|$ less than *or equal to* any given ε, you could take a new $\varepsilon' = \varepsilon/2$ and get

$$|a_n - L| \leq \varepsilon' < \varepsilon,$$

strictly less than ε. Similarly, it suffices to get $|a_n - L|$ less than 3ε, because you can take $\varepsilon' = \varepsilon/3$ and get

$$|a_n - L| < 3\varepsilon' = \varepsilon.$$

So the final ε could be replaced by any constant times ε or anything that's small when ε is small.

3.3. Example of convergence. Prove that $a_n = 1/n^2$ converges to 0.

First, let's think it through. Given $\varepsilon > 0$, we have to see how big n has to be to guarantee that $a_n = 1/n^2$ is within ε of 0:

$$|1/n^2 - 0| < \varepsilon.$$

This will hold if $1/n^2 < \varepsilon$, that is, if $n > 1/\sqrt{\varepsilon}$. So we can just take N to be $1/\sqrt{\varepsilon}$, and we'll have the following proof:

$$\text{Given } \varepsilon > 0, \text{ let } N = 1/\sqrt{\varepsilon}. \text{ Then whenever } n > N,$$

$$|a_n - L| = |1/n^2 - 0| = 1/n^2 < 1/N^2 = \varepsilon.$$

Notice how we had to work backwards to come up with the proof.

3.4. Bounded. A sequence a_n is *bounded* if there is a number M such that for all n, $|a_n| \leq M$.

For example, the sequence $a_n = \sin n$ is bounded by 1 (and by any $M \geq 1$). The sequence $a_n = (-1)^n/n^2$ is bounded by 1. The sequence $a_n = n^2$ is not bounded.

3.5. Proposition. *Suppose that the sequence a_n converges. Then*

(1) *the limit is unique;*

(2) *the sequence is bounded.*

Before starting the proof of (1), let's think about why a sequence cannot have two limits, 0 and 1/4 for example. It's easy for the terms a_n to get within 1 of both, or to get within 1/2 of both, but no better than within $\varepsilon = 1/8$ of both (see Figure 3.1).

Similarly, if a sequence had any two different limits $L_1 < L_2$, you should get a contradiction when $\varepsilon = (1/2)(L_2 - L_1)$. I think I'm ready to write the proof.

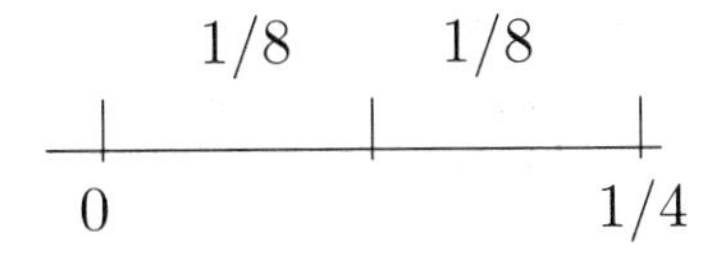

Figure 3.1. A number cannot be closer than distance 1/8 to both 0 and 1/4.

Proof of (1). Suppose that a sequence a_n converges to two different limits $L_1 < L_2$. Let $\varepsilon = (1/2)(L_2 - L_1)$. By the definition of convergence, there is some N_1 such that whenever $n > N_1$, $|a_n - L_1| < \varepsilon$. Similarly there is some N_2 such that whenever $n > N_2$, $|a_n - L_2| < \varepsilon$. Choose n greater than N_1 and N_2. Then

$$L_2 - L_1 \leq |a_n - L_1| + |L_2 - a_n| < \varepsilon + \varepsilon = 2\varepsilon = L_2 - L_1,$$

a contradiction. □

The proof of (2) is easier. After a while the sequence is close to its limit L, and once a_n is within say 1 of L, $|a_n| < |L| + 1$. There are only finitely many other terms to worry about, and of course any finite set is bounded (by its largest element). I'm ready to write the proof:

Proof of (2). Let a_n be a sequence converging to L. Choose N such that whenever $n > N$, $|a_n - L| < 1$, so that $|a_n| < |L| + 1$. Let

$$M = \max\{|L| + 1, |a_n|, \text{ with } n \leq N\}.$$

Then if $n \leq N$, $a_n \leq M$. If $n > N$, $a_n < |L| + 1 \leq M$. So always $a_n \leq M$. □

3.6. Proposition. *Suppose real sequences a_n, b_n converge to a and b:*

$$a_n \to a, \quad b_n \to b.$$

Then

(1) $ca_n \to ca$,

(2) $a_n + b_n \to a + b$,

(3) $a_n b_n \to ab$,

(4) $a_n/b_n \to a/b$, *assuming every b_n and b is a nonzero real number.*

PREPARATION FOR PROOF. We'll prove (1) and (4), and leave (2) and (3) as Exercises 13 and 14. As usual, it pays to build the proof backwards. At the end of the proof of (1), we'll need to estimate

$$|ca_n - ca| = |c|\,|a_n - a| < \varepsilon,$$

which will hold if $|a_n - a| < \varepsilon/|c|$ (unless $c = 0$). I see how to do the proof.

Proof. We may assume that $c \neq 0$, since that case is trivial (it just says that $0, 0, 0, \ldots \to 0$). Since c is a fixed constant, given $\varepsilon > 0$, since $a_n \to a$, we can choose N such that whenever $n > N$, $|a_n - a| < \varepsilon/|c|$. Then

$$|ca_n - ca| = |c|\,|a_n - a| < \varepsilon,$$

so that $ca_n \to ca$.

The proof of (4) is harder, so we start with a discussion. At the end of the proof, we'll need to estimate $|a_n/b_n - a/b|$ in terms of things we know are small: $|a_n - a|$ and $|b_n - b|$. The trick is an old one that you first see in the proof of the quotient rule in calculus: go from a_n/b_n to a/b in two steps, changing one part at a time, from a_n/b_n to a/b_n to a/b, to end up with some estimate like:

$$\begin{aligned} |a_n/b_n - a/b| &\leq |a_n/b_n - a/b_n| + |a/b_n - a/b| \\ &= |a_n - a|/|b_n| + |b - b_n|\,|a/bb_n|. \end{aligned}$$

We know that $|a_n - a|$ and $|b - b_n|$ are small, and $|a/b|$ is just a constant, but what about those $1/|b_n|$? We need to know that b_n is not too close to 0. Fortunately since $b_n \to b$, eventually $|b_n| > |b|/2$ (as soon as b_n gets within $|b|/2$ of b). Then $1/|b_n| \leq 2/|b|$. So the estimate can continue

$$\begin{aligned} &\leq |a_n - a|(2/|b|) + |b - b_n|\,|2a/b^2| \\ &< \varepsilon/2 + \varepsilon/2 = \varepsilon, \end{aligned}$$

if we just make sure that $|a_n - a|(2/|b|)$ and $|b - b_n|\,|2a/b^2|$ are less than $\varepsilon/2$ by making $|a_n - a| < \varepsilon|b|/4$ and $|b - b_n| < \varepsilon|b^2/4a|$ (which we interpret as no condition on b_n if $a = 0$). Can we guarantee both of those conditions at the same time? We can find an N_1 to make the first one hold for $n > N_1$, and we can find an N_2 to make the second one hold for $n > N_2$. To make both work, just take N to be the maximum of N_1 and N_2. In general, you can always handle finitely many conditions.

Here's the whole proof from start to finish. Since $a_n \to a$ and $b_n \to b$, we can choose N such that whenever $n > N$, the following hold:

$$\begin{aligned} &|a_n - a| < \varepsilon|b|/4, \\ &|b - b_n| < \varepsilon|b^2/4a|, \quad \text{and} \\ &|b - b_n| < |b|/2, \quad \text{which implies that } 1/|b_n| \leq 2/|b|. \end{aligned}$$

Then

$$\begin{aligned} |a_n/b_n - a/b| &\leq |a_n/b_n - a/b_n| + |a/b_n - a/b| \\ &= |a_n - a|/|b_n| + |b - b_n|\,|a/bb_n| \\ &\leq |a_n - a|(2/|b|) + |b - b_n|\,|2a/b^2| \\ &< \varepsilon/2 + \varepsilon/2 = \varepsilon, \end{aligned}$$

and consequently $a_n/b_n \to a/b$. □

It would have been OK and simpler to start out by just requiring that

$$\begin{aligned} &|a_n - a| < \varepsilon, \\ &|b - b_n| < \varepsilon, \quad \text{and} \\ &|b - b_n| < |b|/2, \quad \text{which implies that} \quad 1/|b_n| \leq 2/|b|. \end{aligned}$$

Then

$$\begin{aligned}|a_n/b_n - a/b| &\le |a_n/b_n - a/b_n| + |a/b_n - a/b| \\ &= |a_n - a|/|b_n| + |b - b_n|\,|a/bb_n| \\ &< \varepsilon(2/|b|) + \varepsilon|2a/b^2| = C\varepsilon,\end{aligned}$$

where C is the constant $(2/|b|)+|2a/b^2|$. Although we haven't made it come out quite as neatly at the end, we've still shown that we can make the error arbitrarily small by choosing n large enough, which is sufficient.

3.7. Rates of growth. Limits of many sequences can be determined just by knowing that for n large,

$$1/n^2 \ll 1/n \ll 1 \ll \ln n \ll \sqrt{n} \ll n \ll n^2 \ll n^3 \ll 2^n \ll e^n \ll 10^n \ll n!$$

where $f(n) \ll g(n)$ means that f becomes a negligible percentage of g, $f(n)/g(n) \to 0$, so that in a limit as $n \to \infty$ whenever you see $f+g$, or even $c_1 f + c_2 g$, you can ignore the negligible f. For example,

$$\lim_{n\to\infty} \frac{n^4 + \ln(n+1)}{\sqrt{5n^8+16}} = \lim_{n\to\infty} \frac{n^4}{\sqrt{5n^8}} = \frac{1}{\sqrt{5}}.$$

3.8. Three famous limits.

(1) $$\sqrt[n]{2} = 2^{1/n} \to 2^0 = 1.$$

(2) $$\sqrt[n]{n} = n^{1/n} = (e^{\ln n})^{1/n} = e^{(\ln n)/n} \to e^0 = 1.$$

(The exponent $(\ln n)/n \to 0$ because $\ln n \ll n$.)

(3) $$(1+1/n)^n \to e.$$

(This is sometimes used as the definition of the number e, after you check that the limit exists. Exercise 21.6 derives it from another definition.)

3.9. Accumulation points. A point p is an *accumulation point* of a set S if it is the limit of a sequence of points of $S - \{p\}$. It is equivalent to require that every ball $(p-r, p+r)$ about p intersect $S - \{p\}$ (Exercise 20).

For example, 0 is an accumulation point of $\{1/n\colon n \in \mathbb{N}\}$ because $0 = \lim 1/n$ or because every ball (interval) about 0 intersects $\{1/n\}$. In this case the accumulation point is not in the original set.

Every point of the unit interval $[0,1]$ is an accumulation point. In this case all of the accumulation points are in the set.

3.10. $\mathbb{R}^n$. Almost everything in this chapter works for vectors in $\mathbb{R}^n$ as well as for points in $\mathbb{R}$. The exception is Proposition 3.6(4) because there is no way to divide by vectors. Proposition 3.6(3) holds both for the dot product and for the cross product.

Exercises 3

Does the sequence converge or diverge? If it converges, what is the limit?

1. 1, 0, 1/2, 0, 1/4, 0, 1/8, 0, ...

2. 3, 3.1, 3.14, 3.141, 3.1415, 3.14159, ...

3. $a_n = 1 + (-1)^n/n$.

4. $a_n = \frac{1+(-1)^n}{n}$.

5. $a_n = (-1)^n(1 - 1/n)$.

6. $a_n = 1 + (-1)^n$.

7. $a_n = \frac{2n^2+5n+1}{7n^2+4n+3}$.

8. $a_n = \frac{e^n}{n^5+n-5}$.

9. $a_n = \frac{2^n}{n!}$.

10. $a_n = \frac{\sin n}{n}$.

11. Prove that $a_n = 1/n$ converges to 0.

12. Prove that $a_n = 1000/n^3$ converges to 0.

13. Prove 3.6(2).

14. Prove 3.6(3).

15. Prove that if $a_n \leq b_n \leq c_n$ and $\lim a_n = \lim c_n = L$ then $\lim b_n = L$.

16. Prove or give a counterexample. Let a_n be a sequence such that $a_{n+1} - a_n \to 0$. Does a_n have to converge?

17. A sequence a_n is called *Cauchy* if, given $\varepsilon > 0$, there is an N such that whenever $m, n > N$, $|a_m - a_n| < \varepsilon$. Prove that if a sequence in $\mathbb{R}$ is convergent, then it is Cauchy. (Exercise 4.6 will prove the converse in $\mathbb{R}$, so that Cauchy gives a nice criterion for convergence without mentioning what the limit is.)

18. Prove that a Cauchy sequence is bounded.

19. Describe the set of accumulation points of the following sets:

a. the rationals;

b. the irrationals;

c. $[a, b)$;

d. the integers.

20. Prove that p is an accumulation point of S if and only if every ball B about p intersects $S - \{p\}$.

21. Prove or give a counterexample: There are only countably many sequences with limit 0.

22. Prove or give a counterexample: a real increasing sequence

$$a_1 < a_2 < a_3 < \cdots$$

converges if and only if the differences $a_{n+1} - a_n$ converge to 0.

Chapter 4

Subsequences

We like sequences to converge, but most don't. Fortunately, most sequences do have subsequences which converge.

4.1. Discussion. Limits are a kind of ideal, and in general we want sequences that converge. It's a way to find maxima in calculus and in life. Sequences don't always converge, but there's still hope. The sequence

$$1\frac{1}{2},\ -1\frac{1}{2},\ 1\frac{1}{4},\ -1\frac{1}{4},\ 1\frac{1}{8},\ -1\frac{1}{8},\ 1\frac{1}{16},\ -1\frac{1}{16},\ \ldots \tag{4}$$

does not converge, but the odd terms converge to 1, and the even terms converge to -1. These are called convergent subsequences, and these are good things. Given a sequence, a subsequence is some of the terms in the same order. One subsequence of

$$a_1,\ a_2,\ a_3,\ a_4,\ a_5,\ a_6,\ \ldots \tag{5}$$

would be

$$a_2,\ a_3,\ a_5,\ a_7,\ a_{11},\ \ldots, \tag{6}$$

obtained by using just some of the subscripts. The right way to say this, although it may seem a bit confusing at first, is that given a sequence a_n, we can consider a subsequence a_{m_n}, with $m_1 < m_2 < m_3 < \cdots$, where above $m_1 = 2$, $m_2 = 3$, $m_3 = 5$, $m_4 = 7$, $m_5 = 11$, ...

4.2. Definition of subsequence. Given a sequence a_n, a *subsequence* a_{m_n} consists of some of the terms in the same order.

Now given a sequence, what is the chance that it has a convergent subsequence? Of course if it is convergent, every subsequence, including the original sequence, converges to the same limit. On the other hand, a divergent sequence like

(1) $$1,\ 2,\ 3,\ 4,\ \ldots$$

has no convergent subsequence because every subsequence diverges to ∞. The following is a delightful and important theorem due to Bolzano and Weierstrass, around 1840.

4.3. Theorem. *Every bounded sequence in $\mathbb{R}$ has a convergent subsequence.*

Proof. We'll first do the case of nonnegative sequences and then treat the general case. So we consider a nonnegative sequence

$$a_1,\ a_2,\ a_3,\ \ldots.$$

Each a_n starts off with a nonnegative integer before the decimal point, followed by infinitely many digits (possibly 0) after the decimal point. Since the sequence a_n is bounded, some integer part D before the decimal place occurs infinitely many times. Throw away the rest of the a_n. Among the infinitely many remaining a_n that start with D, some first decimal place d_1 occurs infinitely many times. Throw away the rest of the a_n. Among the infinitely many remaining a_n that start with $D.d_1$, some second decimal place d_2 occurs infinitely many times. Throw away the rest of the a_n. Among the infinitely many a_n that start with $D.d_1d_2$, some third decimal place d_3 occurs infinitely many times. Keep going to construct $L = D.d_1d_2d_3\ldots$.

We claim that there is a subsequence converging to L. Let a_{m_1} be one of the infinitely many a_n which starts with $D.d_1$. Let a_{m_2} be a later a_n which starts with $D.d_1d_2$. Let a_{m_3} be a later a_n which starts with $D.d_1d_2d_3$. The a_{m_n} converge to L because once $n > N$ they agree with L for at least N decimal places.

Having treated the case of nonnegative sequences, we now consider a sequence with some negative terms. Since the sequence is bounded, we can translate it to the right to make it nonnegative, use the above argument to get a convergent subsequence, and then translate it back. □

4.4. Corollary. *An increasing sequence in $\mathbb{R}$ which is bounded above converges.*

(Similarly, a decreasing sequence bounded below converges. The sequence need not be *strictly* increasing or decreasing.)

Proof. Exercise 3. □

Exercises 4

1. Give an example of a sequence of real numbers with two different subsequences converging to two different limits.

2. Give an example of a sequence of real numbers with subsequences converging to each integer.

3. Prove Corollary 4.4.

4. Give an example of a sequence of real numbers with subsequences converging to every real number.

5. Prove that a set $S \subset \mathbb{R}$ is bounded if and only if every sequence of points in S has a convergent subsequence.

6. Prove that every Cauchy sequence of reals converges. (Use Exercises 3.17 and 3.18.)

7. The *lim sup* of a sequence is defined as the largest limit of any subsequence, or $\pm\infty$. (It always exists.) What is the lim sup of the following sequences?

a. $a_n = (-1)^n$

b. $1, -1, 2, -2, 3, -3, \ldots$

c. $a_n = -n^2$

d. $a_n = \sin(n\pi/6)$

e. $\frac{1}{2}, 1\frac{1}{2}, 2\frac{1}{2}, \frac{1}{4}, 1\frac{1}{4}, 2\frac{1}{4}, \frac{1}{8}, 1\frac{1}{8}, 2\frac{1}{8}, \ldots$

f. $\{a_n\} = \mathbb{Q} \cap (0, 1)$

8. Similarly the *lim inf* of a sequence is defined as the smallest limit of any subsequence, or $\pm\infty$. Of course the lim inf is less than or equal to the lim sup. Give two examples of sequences for which the lim inf equals the lim sup.

Chapter 5

Functions and Limits

Recall that for us a function f takes an input x from some domain $E \subset \mathbb{R}^n$ and produces an output $f(x) \in \mathbb{R}$. For example, for $f(x) = 1/x$, the natural domain is $E = \mathbb{R} - \{0\}$.

We want to extend our notion of limit from sequences to functions. Instead of the limit $\lim_{n\to\infty} a_n$ of a sequence, such as $\lim_{n\to\infty} 1/n = 0$, we want to consider a limit $\lim_{x\to p} f(x)$ of a function, such as $\lim_{x\to 3} x^2 = 9$, to give a trivial example. Instead of going way out in our sequence $(n > N)$, we will be taking x close to some fixed target p, taking $|x - p|$ small. To say how small, we'll use the Greek letter before epsilon ε, namely delta δ, and require that $|x - p| < \delta$.

5.1. Definitions. We say $\lim_{x\to p} f(x) = a$ ("the limit as x approaches p of $f(x)$ equals a") if, given $\varepsilon > 0$, there exists a $\delta > 0$ such that

$$0 < |x - p| < \delta \Rightarrow |f(x) - a| < \varepsilon.$$

Of course we have to assume that x lies in the domain E of f, and hence we need to assume that p is an accumulation point of E. Notice that by requiring that $0 < |x - p|$, we do not allow x to equal p. The limit depends only on values $f(x)$ at nearby points, not on $f(p)$. If it happens that $\lim_{x\to p} f(x) = f(p)$, or p is not an accumulation point of E, then f is called *continuous* at p. In other words, f is continuous at p if given $\varepsilon > 0$, there exists $\delta > 0$, such that

$$|x - p| < \delta \implies |f(x) - f(p)| < \varepsilon.$$

(Here it doesn't matter whether we allow x to equal p, since of course when x equals p, $f(x) = f(p)$ and the right hand side holds automatically.)

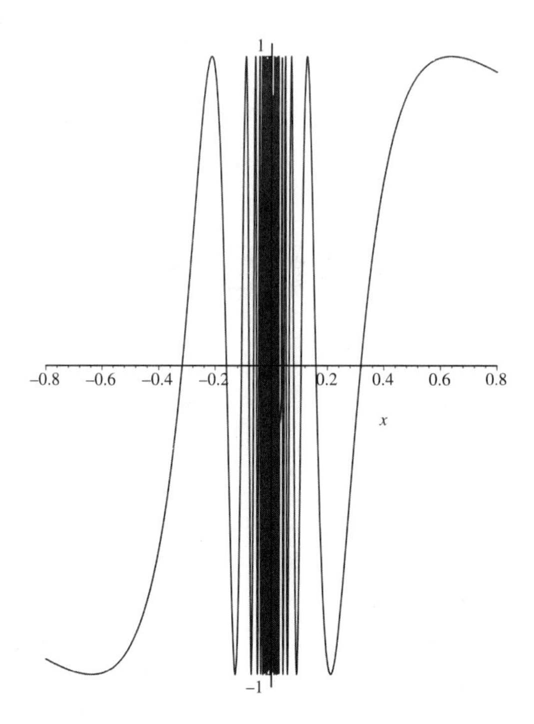

Figure 5.1. For $f(x) = \sin \frac{1}{x}$, $\lim_{x\to 0} f(x)$ does not exist.

Examples of continuous functions include powers of x, $\sin x$, $\cos x$, e^x, and combinations of continuous functions such as

$$\frac{e^x \cos x}{x^2 + 1}$$

(as long as the denominator is never 0).

5.2. Example. Consider the function $f(x) = \sin \frac{1}{x}$, defined on $E = \mathbb{R} - \{0\}$, graphed in Figure 5.1. Then $\lim_{x\to 0} f(x)$ does not exist (Exercise 4).

5.3. Example. Consider the function $f(x) = x \sin \frac{1}{x}$, defined on $E = \mathbb{R} - \{0\}$, graphed in Figure 5.2. Then $\lim_{x\to 0} f(x) = 0$ (Exercise 3). If we define $f(0) = 0$, then f will be continuous.

5.4. Example. Consider the function $f(x)$ on $\mathbb{R}$ which is 1 on rationals and 0 on irrationals. Then $\lim f(x)$ never exists, and f is continuous nowhere. This function is called the *characteristic function* of the rationals and written

$$f = \chi_{\mathbb{Q}},$$

using the Greek letter chi.

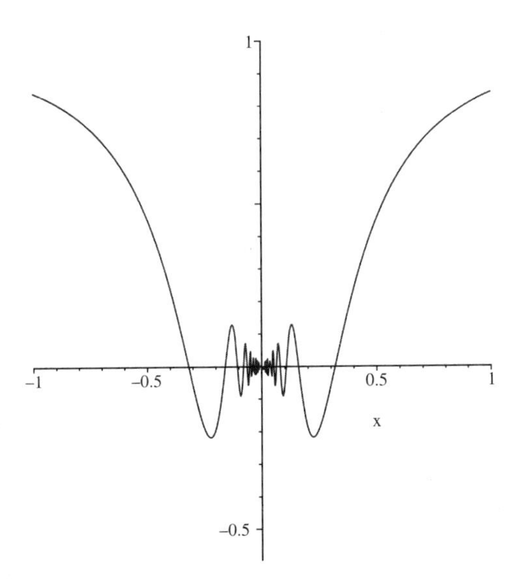

Figure 5.2. For $f(x) = x \sin \frac{1}{x}$, $\lim_{x \to 0} f(x) = 0$.

5.5. Proposition. *If* $\lim_{x \to p} f(x)$ *exists, then it is unique.*

Proof. The proof is just like the proof that the limit of a sequence is unique. Suppose that $\lim_{x \to p} f(x)$ has two distinct values a, b. Choose $\delta > 0$ such that if $0 < |x - p| < \delta$, then

$$|f(x) - a| < |b - a|/2 \quad \text{and} \quad |f(x) - b| < |b - a|/2.$$

Since for limits we always assume that p is an accumulation point of the domain, some x_0 satisfies $0 < |x_0 - p| < \delta$. Then

$$|b - a| \leq |f(x_0) - a| + |f(x_0) - b| < |b - a|/2 + |b - a|/2 = |b - a|,$$

the desired contradiction. □

5.6. Proposition. *Suppose that* $\lim_{x \to p} f(x) = a$ *and* $\lim_{x \to p} g(x) = b$. *Then*

(1) $\lim_{x \to p} cf(x) = ca$,

(2) $\lim_{x \to p}(f + g)(x) = a + b$,

(3) $\lim_{x \to p}(fg)(x) = ab$, *and*

(4) $\lim_{x \to p}(f/g)(x) = a/b$ *if* $b \neq 0$.

Proof. The proof is similar to the proof of 3.6; see Exercises 5 and 6. □

5.7. Proposition. *A function f is continuous at p if and only if for every sequence $x_n \to p$, $f(x_n) \to f(p)$.*

Proof. Suppose that f is continuous at p. Then *all* points x near p have values $f(x)$ near $f(p)$, so certainly the x_n for n large will have values $f(x_n)$ near $f(p)$. Conversely, suppose that f is not continuous at p. That means that for some $\varepsilon > 0$, no small δ bound on $|x - p|$, for example $\delta = 1/n$, guarantees that $|f(x) - f(p)| \leq \varepsilon$. Thus there must be some sequence x_n, with $|x_n - p| \leq 1/n$, with $|f(x_n) - f(p)| > \varepsilon$, so that $f(x_n)$ does not converge to $f(p)$. □

Exercises 5

1. Give an example of a function $f\colon \mathbb{R} \to \mathbb{R}$ which is continuous except at the integers.

2. Give an example of a function $f\colon \mathbb{R} \to \mathbb{R}$ which is continuous only at 0.

3. Prove that for the function $f(x)$ of Example 5.3, $\lim_{x \to 0} f(x) = 0$.

4. Prove that for the function $f(x)$ of Example 5.2, $\lim_{x \to 0} f(x)$ does not equal 0.

5. Prove 5.6(2).

6. Prove 5.6(3).

7. Prove that the sum of two continuous functions is continuous.

8. Prove that the product of two continuous functions is continuous.

9. Give an example of a function which is continuous only at the integers. Graph it.

10. Give an example of two functions, both discontinuous at 0, whose sum is continuous at 0. Give an example of two functions, both discontinuous at 0, whose product is continuous at 0.

11. Consider the function $f\colon \mathbb{R} \to \mathbb{R}$ which is 0 at irrationals and $1/q$ at a rational p/q (in lowest terms with q positive). Where is f continuous?

12. Prove that every function $f\colon \mathbb{Z} \to \mathbb{R}$ is continuous, first from the ε-δ definition of continuous, second using Proposition 5.7.

Chapter 6

Composition of Functions

Composition is an important way to build more complicated functions out of simpler ones. Thus you can understand complicated functions by analyzing simpler ones.

6.1. Definition. The *composition* $f \circ g$ of two functions is the function obtained by following one with the other:

$$(f \circ g)(x) = f(g(x)).$$

For this to make sense, the image of g must be contained in the domain of f.

Examples. If $f(x) = \sin x$ and $g(x) = x^2$, then

$$(f \circ g)(x) = \sin(x^2), \quad \text{usually written simply } \sin x^2;$$
$$(g \circ f)(x) = (\sin x)^2, \quad \text{usually written } \sin^2 x.$$

If $f(x) = \sqrt{x}$ and $g(x, y) = x^2 + y^2$, then

$$(f \circ g)(x) = \sqrt{x^2 + y^2}.$$

6.2. Theorem. *The composition $f \circ g$ of two continuous functions is continuous.*

Proof. We will give two proofs, the first from the ε-δ definition of continuous, and a second, slicker proof using the sequence characterization of continuous (Proposition 5.7). See Figure 6.1.

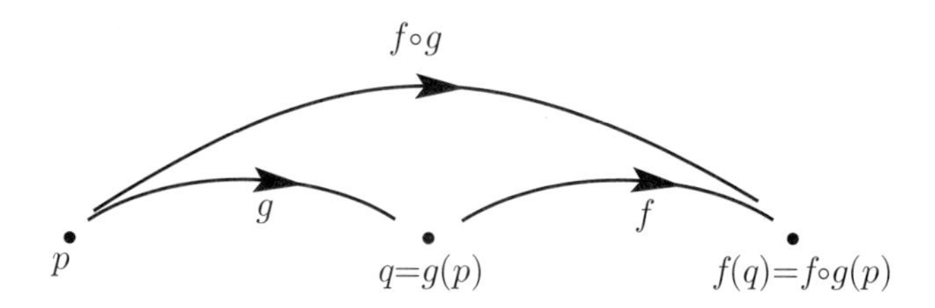

Figure 6.1. The composition $f \circ g$ of two functions.

ε-δ definition proof. Given a point p in the domain of g, let $q = g(p)$ be the resulting point in the domain of f, so that $(f \circ g)(p) = f(q)$. Given $\varepsilon > 0$, since f is continuous, we can choose $\varepsilon_1 > 0$ such that

$$|y - q| < \varepsilon_1 \Rightarrow |f(y) - f(q)| < \varepsilon. \tag{2}$$

Since g is continuous, we can choose $\delta > 0$ such that

$$|x - p| < \delta \Rightarrow |g(x) - g(p)| < \varepsilon_1.$$

Now applying (2) with $y = g(x)$ and $q = g(p)$ implies that

$$|f(g(x)) - f(g(p))| < \varepsilon,$$

as desired.

Sequence definition proof. Suppose $x_n \to x$. Since g is continuous, $g(x_n) \to g(x)$. Then since f is continuous, $f(g(x_n)) \to f(g(x))$, as desired. □

Exercises 6

1. What is $f \circ g(x)$ if

a. $f(x) = \sqrt{x}$ and $g(x) = x^2$;

b. $f(x) = \sin x$ and $g(x) = \sin^{-1} x$.

2. Show by counterexample that this converse of Theorem 6.2 is false: If the composition of two functions is continuous, then both are continuous.

Part II

Topology

Chapter 7

Open and Closed Sets

Although sequences played the crucial role in the rigorous understanding of calculus developed in the 1800s, over the next fifty years mathematicians sharpened the theory using the concepts of open and closed sets. If it took mathematicians fifty years to get used to these more abstract ideas, you shouldn't worry if it takes you a couple of weeks. It's actually quite fun if you take your time and think about lots of examples.

7.1. Definition. A point p in $\mathbb{R}^n$ is a *boundary* point of a set S in $\mathbb{R}^n$ if every ball about p meets both S and its complement S^{C}. The set of boundary points of S is called the boundary of S and written ∂S. (That funny symbol, sometimes called *del*, is not a Greek letter. It is used for partial differentiation too.)

It follows immediately that a set S and its complement S^{C} have the same boundary.

For example, consider a *ball* about a point a of radius $r > 0$:

$$B(a, r) = \{|x - a| \leq r\}$$

(see Figures 7.1 and 7.2). Its boundary is the sphere:

$$\partial B(a, r) = \{|x - a| = r\}.$$

In this case the boundary of S is part of S. In $\mathbb{R}^2$, we sometimes call the ball a *disc* and we usually call its boundary a *circle* rather than a sphere. In $\mathbb{R}^1$, the ball $B(a, r)$ is just the interval $[a - r, a + r]$ and its boundary is just the two endpoints.

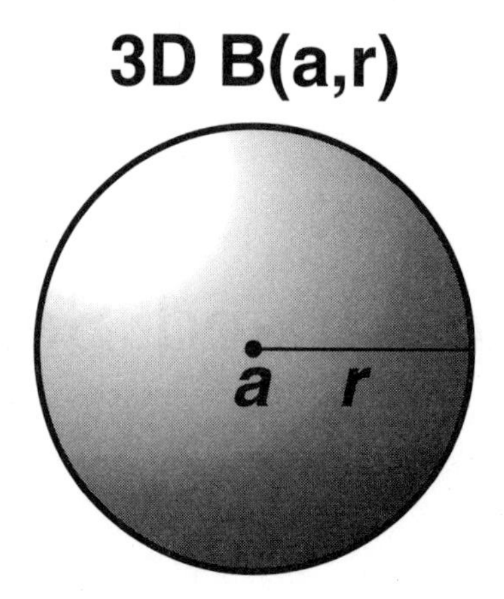

Figure 7.1. The boundary of the ball $B(a, r)$ is the sphere. The closed ball includes its boundary.

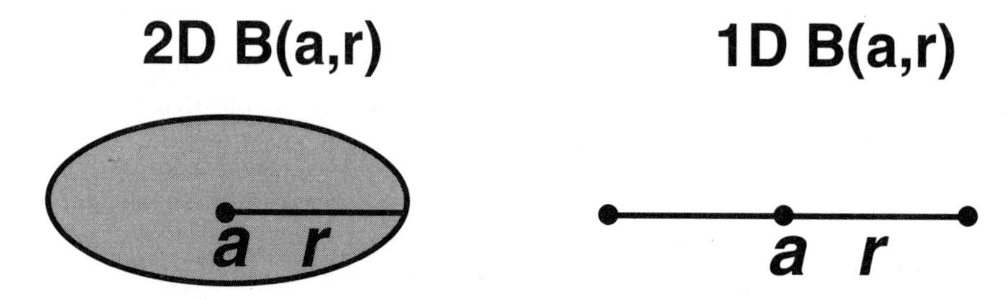

Figure 7.2. The boundary of a 2D ball or "disc" is a circle. The boundary of a 1D ball or interval is two points.

As another example, consider all of $\mathbb{R}^n$ with one point, the origin, removed:

$$S = \mathbb{R}^n - \{0\}.$$

Its boundary is the single point $\{0\}$. In this case, the boundary of S is not part of S.

The boundary of the rationals $\mathbb{Q}$ is all of $\mathbb{R}$. The boundary of $\mathbb{R}$ is the empty set.

7.2. Definition. A set S in $\mathbb{R}^n$ is *open* if it contains none of its boundary points. A set S in $\mathbb{R}^n$ is *closed* if it contains all of its boundary points.

It follows immediately that a set S is open if and only if its complement $S^{\complement}$ is closed.

Many a set contains just part of the boundary and hence is neither open nor closed. "Not closed" does not mean "open." These terms are not opposites!

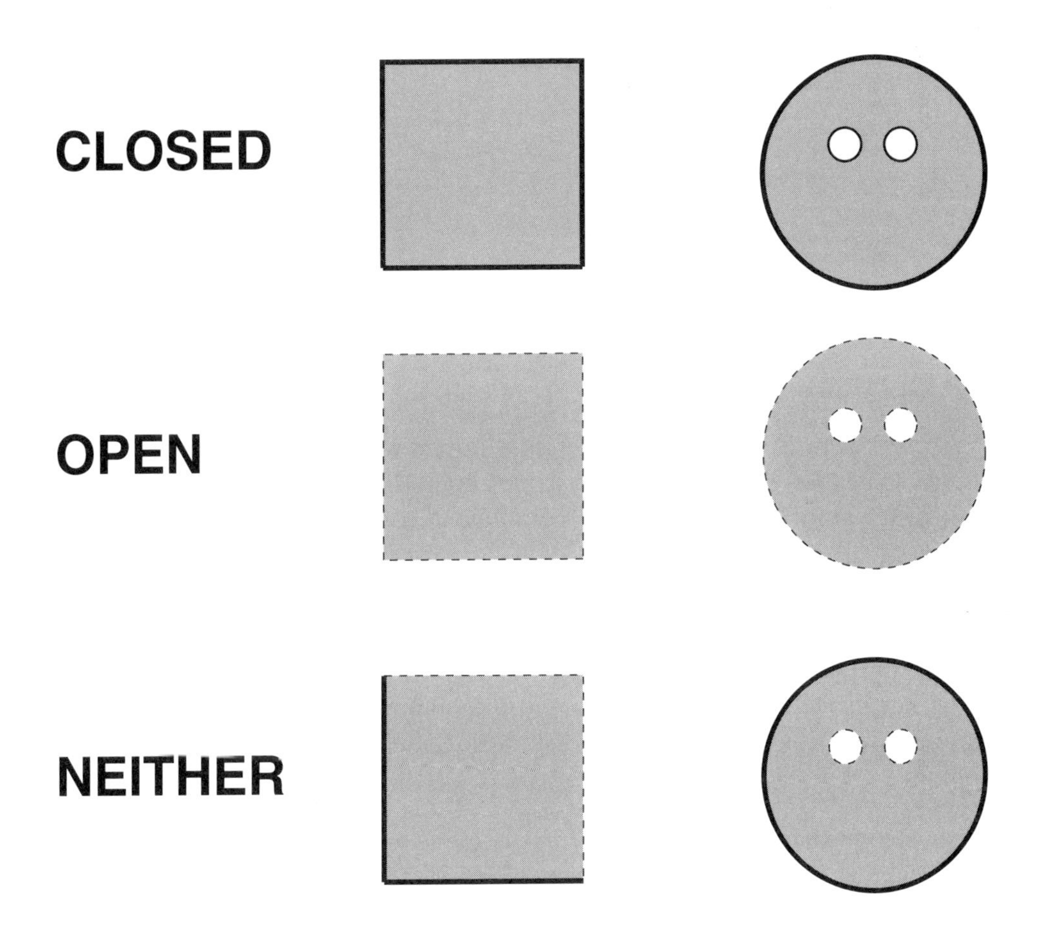

Figure 7.3. Some sets are closed or open but most are neither.

For example, the ball $B(a, r)$ is closed. If you remove its boundary, the resulting set is open and is called an open ball, for which unfortunately we have no special symbol. $\mathbb{R}^n - \{0\}$ is open. The rationals $\mathbb{Q}$ are neither open nor closed. The reals $\mathbb{R}$ are both open and closed. See Figure 7.3 for a few more examples.

Proposition 7.3 gives nice direct characterizations of open and closed. A set is open if it includes a ball about every point; of course the ball has to get smaller as you get close to the boundary. A set is closed if it includes all accumulation points. See Figure 7.4.

7.3. Proposition. *A set S in $\mathbb{R}^n$ is open if and only if about every point of S there is a ball completely contained in S. A set S is closed if and only if it contains all of its accumulation points.*

Proof. Suppose that about some point p of S there is no ball completely contained in S. Then every ball about p contains a point of S^{C} as well as

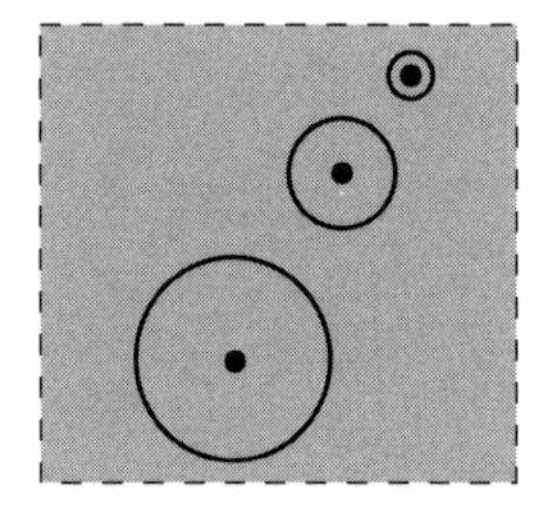

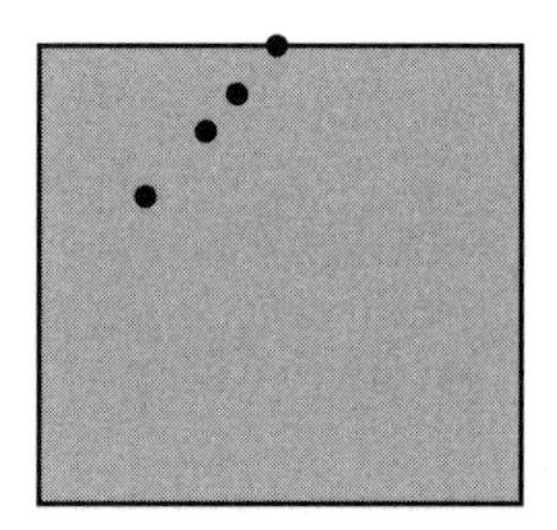

Figure 7.4. An open set includes a ball about every point. A closed set includes all of its accumulation points.

the point p in S. Consequently, p is a boundary point of S, and S is not open. Conversely, suppose that S is not open. Then some point p of S is a boundary point, and there is no ball about p contained in S.

Suppose that an accumulation point p of S is not contained in S. Then every ball about p contains a point of $S^{\complement}$ (namely, p) and a point of S (because p is an accumulation point of S). Therefore, p is a boundary point not contained in S, so that S is not closed. Conversely, suppose that S is not closed. Then $S^{\complement}$ contains some boundary point p of S, which is an accumulation point of S, so S does not contain all of its accumulation points. □

7.4. Proposition. *Any union of open sets is open. A finite intersection of open sets is open. Any intersection of closed sets is closed. A finite union of closed sets is closed.*

Proof. To prove the first statement, suppose that x belongs to the union of open sets U_α. Then x belongs to some U_β. By Proposition 7.3, some ball about x is contained in U_β, and hence in the union of all the U_α. Therefore, by Proposition 7.3 again, the union is open.

To prove the second statement, let U be the intersection of finitely many open sets U_i. Let p belong to U. Since p belongs to U_i, there is a ball $B(p, r_i)$ contained in U_i. Let $r = \min\{r_i\}$. Then $B(p, r)$ is contained in every U_i and hence in U. Therefore U is open.

Let C be an intersection of closed sets C_α. Then $C^{\complement}$ is the union of the open sets $C_\alpha^{\complement}$, and hence open. Therefore C is closed. Similarly let C be a union of finitely many closed sets C_i. Then $C^{\complement}$ is the intersection of the open sets $C_i^{\complement}$, and hence open. Therefore C is closed. □

7.5. Definitions (see Figure 7.5). The *interior* of a set S, denoted int S or $\mathring{S}$, is $S - \partial S$. The *closure* of S, denoted cl S or $\bar{S}$, is $S \cup \partial S$. An *isolated point* of S is the only point of S in some ball about it.

Figure 7.5. A set, its interior, its closure, and an isolated point p.

7.6. Proposition. *The interior of S is the largest open set contained in S, and the closure of S is the smallest closed set containing S.*

Proof. Exercises 11 and 12. □

7.7. Topology. In $\mathbb{R}^n$ or in more general spaces, the collection of open sets is called the *topology*, which determines, as we'll soon see, continuity, compactness, connectedness, and other "topological" properties.

Exercises 7

1. Say whether the following subsets of $\mathbb{R}$ are open, closed, neither, or both. Give reasons.

a. $[0, 1)$;

b. $\mathbb{Z}$;

c. $\{x \in \mathbb{R}\colon \sin x > 0\}$,

d. $\bigcup_{n=2}^{\infty} [1/n, 1)$.

2. Show by example that the intersection of infinitely many open sets need not be open.

3. Show by example that the union of infinitely many closed sets need not be closed.

4. Is all of $\mathbb{R}$ the only open set containing $\mathbb{Q}$? Prove your answer correct.

5. If $S = [0, 1)$, what are ∂S, $\mathring{S}$, and $\bar{S}$?

6. If $S = \mathbb{Z}$, what are ∂S, $\mathring{S}$, and $\bar{S}$?

7. If $S = B(0,1)$ in $\mathbb{R}^2$, what are ∂S, $\mathring{S}$, and $\bar{S}$?

8. Give an example of open sets $U_1 \supset U_2 \supset \cdots$ such that the intersection $\bigcap U_i$ is closed and nonempty.

9. Give an example of nonempty closed sets $C_1 \supset C_2 \supset \cdots$ such that the intersection $\bigcap C_i$ is empty.

10. Prove that every point of S is either an interior point or a boundary point.

11. Prove that the interior of S is the largest open set contained in S.

12. Prove that the closure of S is the smallest closed set containing S.

13. Prove that a boundary point of S is either an isolated point or an accumulation point.

14. Prove or give a counterexample: two disjoint sets cannot each be contained in the other's boundary.

15. A subset S_0 of a set S is *dense* in S if every ball about every point of S contains a point of S_0. Are the rationals dense in the reals? Are rationals with powers of 2 in the denominator dense in the reals? Are the points with rational coordinates dense in $\mathbb{R}^n$?

Chapter 8

Compactness

Probably the most important new idea you'll encounter in real analysis is the concept of compactness. It's the compactness of $[a, b]$ that makes a continuous function reach its maximum and that makes the Riemann integral exist. By definition, compactness means that every sequence has a convergent subsequence. For subsets of $\mathbb{R}^n$, this turns out to be equivalent to *closed* and *bounded.*

8.1. Definitions. Let S be a set in $\mathbb{R}^n$. S is *bounded* if it is contained in some ball $B(0, R)$ about 0. S is *compact* if every sequence in S has a subsequence converging to a point of S.

8.2. Theorem. (Bolzano–Weierstrass). *A set S in $\mathbb{R}^n$ is compact if and only if S is closed and bounded.*

A nonclosed set such as $(0, 1]$ is not compact for example because every subsequence of the sequence $a_n = 1/n$ converges to 0, which is not in $(0, 1]$. An unbounded set such as $\mathbb{R}$ is not compact because the sequence $a_n = n$ has no convergent subsequence. This is the main idea of the first part of the proof.

Proof. Suppose that S is not closed. Let p be an accumulation point not in S, and let a_n be a sequence of points in S converging to p. Then every subsequence converges to p, which is not in S.

Suppose that S is not bounded. Let a_n be a sequence of points with $|a_n|$ diverging to infinity. Then a_n has no convergent subsequence.

Finally suppose that S is closed and bounded. Take any sequence of points in $S \subset \mathbb{R}^n$. First look at just the first of the n components of each point. Since S is bounded, the sequence of first components is bounded. By Theorem 4.3, for some subsequence, the first components converge. Similarly, for some further subsequence, the second components also converge. Eventually, for some subsequence, all of the components converge. Since S is closed, the limit is in S. □

8.3. Proposition. *A nonempty compact set S of real numbers has a largest element (called the* maximum*) and a smallest element (called the* minimum*).*

Proof. We may assume that S has some positive numbers, by translating it to the right if necessary. Since S is bounded, there is a largest integer part D before the decimal place. Among the elements of S that start with D, there is a largest first decimal place d_1. Among the elements of S that start with $D.d_1$, there is a largest second decimal place d_2. Keep going to construct $a = D.d_1d_2d_3\ldots$. By construction, a is in the closure of S. Since S is closed, a lies in S and provides the desired maximum.

A minimum is provided by $-\max(-S)$. □

Exercises 8

1. Prove that the intersection of two compact sets is compact, directly from the definition of compact.

2. Prove that the intersection of two compact sets is compact, using Theorem 8.2.

3. Prove that the intersection of infintely many compact sets is compact.

4. Prove that the union of two compact sets is compact, using Theorem 8.2.

5. Prove that the union of two compact sets is compact, directly from the definition of compact.

6. Is the union of infinitely many compact sets always compact? Give a proof or counterexample.

7. Is the converse of Proposition 8.3 true? Give a proof or counterexample.

8. Prove that if a nonempty closed set S of real numbers is bounded above, then it has a largest element.

9. Define the *supremum* $\sup S$ of a nonempty bounded set of real numbers as $\max \bar{S}$. Prove that $\sup S \geq s$ for all s in S and that $\sup S$ is the smallest number with that property. For this reason, $\sup S$ is often called the *least upper bound*. Similarly, the *infimum* $\inf S = \min \bar{S}$ is called the *greatest lower bound*.

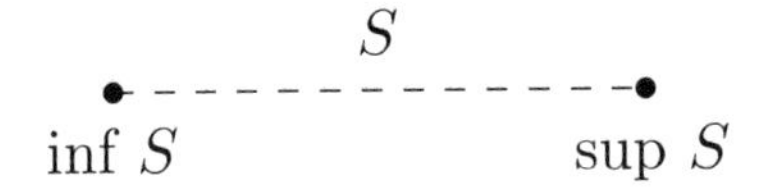

(If S is not bounded above, $\sup S$ is defined to be $+\infty$. If S is empty, $\sup S$ is defined to be $-\infty$.)

Chapter 9

Existence of Maximum

One of the main theorems of calculus is that a continuous function on a bounded, closed interval $[a, b]$ attains a maximum (and a minimum). It is the reason that the calculus method of finding maxima and minima works. It appears as Corollary 9.2 below. As we will now see, this important result depends on the fact that the interval is compact.

9.1. Theorem. *A continuous image of a compact set is compact.*

We will give a proof using the Bolzano–Weierstrass sequential definition of compactness. This is a very general argument, which holds in $\mathbb{R}^n$. We will not give a proof using the "closed and bounded" characterization of compactness because I do not know one. The two natural steps are false, as shown by Exercises 2 and 3. Somehow when the two notions are combined in "compactness," the proof comes faster.

Proof. Let f be a continuous function, let K be a compact set, and let a_n be a sequence in $f(K)$. Choose b_n in K such that $f(b_n) = a_n$. Since K is compact, a subsequence b_{m_n} converges to some limit b in K. Since f is continuous, the corresponding subsequence $a_{m_n} = f(b_{m_n})$ converges to $f(b)$ in $f(K)$. □

9.2. Corollary (Existence of extrema). *A continuous function on a nonempty compact set K, such as a closed bounded interval $[a, b]$, attains a maximum and a minimum.*

Proof. By Theorem 9.1, $f(K)$ is compact. By Proposition 8.3, $f(K)$ has a maximum and a minimum. □

Remark. Corollary 9.2 depends crucially on the compactness of an interval $[a, b]$ of real numbers. Note that it fails for the rationals. For example, on the interval of rationals from 0 to 2, the continuous function $(x^2 - 2)^2$ has no minimum; the continuous function $1/(x^2 - 2)$ is not even bounded.

Exercises 9

1. Give an example of a function (not continuous) on $[0, 1]$ which has no maximum and no minimum.

2. Give an example of a continuous function on a closed (but unbounded) set $A \subset \mathbb{R}$ that has no maximum.

3. Give an example of a continuous function on a bounded (but not closed) set $A \subset \mathbb{R}$ that has no maximum.

4. Prove that if K is compact and f and g are continuous, then $(f \circ g)(K)$ is compact. What do you have to assume about the domains and images of f and g?

5. Prove or give a counterexample: A nonempty set K in $\mathbb{R}$ is compact if and only if every continuous function on K has a maximum.

6. (Candice Corvetti). Prove or give a counterexample: A function $f : \mathbb{R} \to \mathbb{R}$ is continuous if and only if the image of every compact set is compact.

Chapter 10

Uniform Continuity

For integration, it turns out that continuity is not quite strong enough. One needs a kind of "uniform continuity," independent of x.

10.1. Definition. A function f is *uniformly continuous* if, given $\varepsilon > 0$, there is a $\delta > 0$ such that

$$|y - x| < \delta \Rightarrow |f(y) - f(x)| < \varepsilon.$$

The only difference from mere continuity is that the δ appears before the x, and consequently cannot depend on x. The same δ must work for all x.

10.2. Theorem. *A continuous function on a compact set K is uniformly continuous.*

This theorem fails on general sets. For example, $f(x) = x^2$ is continuous on $\mathbb{R}$ but not uniformly continuous. To show that it is not uniformly continuous, we have to come up with an ε for which no δ works. In this case, any ε will do; let's take $\varepsilon = 1$. Now no matter how small δ is, there are large x's within δ of each other for which the values of x^2 differ by more than 1. Exercise 4 asks for the details. Geometrically, the problem is that the slope goes to ∞.

Likewise, $f(x) = 1/x$ is continuous on $(0, 1)$ but not uniformly continuous (Exercise 5).

Proof of Theorem 10.2. Suppose that there is a continuous function f on K which is not uniformly continuous. For some $\varepsilon > 0$, taking $\delta = 1/n$, there are points x_n, y_n such that

$$|y_n - x_n| < 1/n, \quad |f(y_n) - f(x_n)| \geq \varepsilon. \tag{1}$$

Since K is compact, by taking a subsequence we may assume that $x_n \to x$. By taking a further subsequence, we may also assume that $y_n \to y$. By (1) and continuity, $x = y$, but $|f(y) - f(x)| \geq \varepsilon$, a contradiction. □

Exercises 10

1. Give another example of a continuous function on a closed set which is not uniformly continuous.

2. Give another example of a continuous function on a bounded set which is not uniformly continuous.

3. Prove that a composition of uniformly continuous functions is uniformly continuous.

4. Prove that the function $f(x) = x^2$ is not uniformly continuous on $\mathbb{R}$.
Hint: Take $\varepsilon = 1$, any $\delta > 0$, $x = 1/\delta$, $y = 1/\delta + \delta/2$, and show that even though $|y - x| < \delta$, $|f(y) - f(x)| \geq \varepsilon$.

5. Prove that the function $f(x) = 1/x$ is not uniformly continuous on $(0, 1)$.

6. Prove that a uniformly continuous image of a Cauchy sequence is Cauchy. Show by counterexample that *uniformly* is necessary.

7. (Zan Armstrong). Prove or give a counterexample to the following converse of Theorem 10.2. A set $S \subset \mathbb{R}$ is compact if every continuous function on S is uniformly continuous.

Chapter 11

Connected Sets and the Intermediate Value Theorem

The third of the big Cs requisite for calculus, after "continuous" and "compact," is "connected." The famous Intermediate Value Theorem follows easily.

11.1. Definition. A set S is *connected* if it cannot be separated by two disjoint open sets U_1 and U_2 into two nonempty pieces $S \cap U_1$ and $S \cap U_2$ (such that $S = (S \cap U_1) \cup (S \cap U_2)$).

For example, the subset S of the reals defined by

$$S = \{0\} \cup [1, 3]$$

is disconnected, as may be seen by taking $U_1 = (-1/2, 1/2)$, $U_2 = (1/2, 4)$. The rationals $\mathbb{Q}$ are disconnected, as may be seen for example by taking $U_1 = (-\infty, \sqrt{2})$ and $U_2 = (\sqrt{2}, \infty)$. A singleton is obviously connected. Intervals of reals are connected, although that requires proof:

11.2. Proposition. *An interval I of real numbers is connected.*

Proof. Suppose that I can be separated by two disjoint open sets into two nonempty pieces $I \cap U_1$ and $I \cap U_2$. Let $a_i \in I \cap U_i$. We may suppose that $a_1 < a_2$. Let b_1 be the largest element of $[a_1, a_2] - U_2$ (which is compact and nonempty because it has a_1 in it). Let b_2 be the smallest element of $[b_1, a_2] - U_1$ (which is compact and nonempty because it has a_2 in it). Then

$b_1 < b_2$. Choose $b_1 < b_3 < b_2$. By choice of b_2, the point $b_3 \in U_1$ and hence $b_3 \notin U_2$, which contradicts the choice of b_1. □

The converse also is true:

11.3. Proposition. *Every connected subset S of $\mathbb{R}$ is an interval (or a single point or empty).*

Proof. If S is not an interval (or a single point or empty), then there are points a, b in S and a point c between a and b not in S. Then $U_1 = (-\infty, c)$ and $U_2 = (c, \infty)$ show that S is not connected. □

11.4. Theorem. *The continuous image of a connected set is connected.*

Proof. If $f(S)$ is disconnected by U_1, U_2, then S is disconnected by $f^{-1}U_1$, $f^{-1}U_2$ (which are open because f is continuous). □

11.5. The Intermediate Value Theorem. *Let f be a continuous real-valued function on a closed and bounded interval $[a, b]$. Then f attains all values between $f(a)$ and $f(b)$.*

Proof. By 11.2 and 11.4, $f([a, b])$ is connected. By Proposition 11.3, it contains all values between $f(a)$ and $f(b)$. □

11.6. Definition. First note that a set is disconnected if and only if it can be separated by two disjoint open sets U_1 and U_2 into two nonempty pieces $S \cap U_1$ and $S \cap U_2$. A set S is *totally disconnected* if it has at least two points and for all distinct points p_1, p_2 in S, the set S can be separated by two disjoint open sets U_1 and U_2 into two pieces $S \cap U_1$ and $S \cap U_2$ containing p_1 and p_2, respectively.

For example, the integers and the rationals are totally disconnected (Exercise 10). Note that every totally disconnected set is disconnected.

11.7. Proposition. *A set of reals is totally disconnected if and only if it contains at least two points but no interval.*

Proof. Exercise 11. □

Exercises 11

1. Prove from the definition that the integers are disconnected.

2. Give a counterexample to the following statement: If $f : \mathbb{R} \to \mathbb{R}$ is continuous and S is connected, then $f^{-1}S$ is connected.

3. a. Show that $\mathbb{Q} \times \mathbb{Q} \subset \mathbb{R}^2$ is disconnected.
b. Is $\mathbb{Q} \times \mathbb{Q} \subset \mathbb{R}^2$ totally disconnected?

4. Is $[0, 1) \times [0, 1) \subset \mathbb{R}^2$ connected?

5. Is the unit circle $\{x^2 + y^2 = 1\}$ in $\mathbb{R}^2$ connected?

6. Prove that your height (in inches) once equaled your weight (in pounds).

7. Prove that if f is a continuous function from $\mathbb{R}$ to $\mathbb{Q}$, then f is constant.

8. What can you say about the continuous image of an interval $[a, b]$?

9. Let f be a continuous function from $\mathbb{R}$ to $\mathbb{R}$ such that $f(x)^2 = x^2$. What are the possibilities for f? Prove your answer correct.

10. Prove that the integers $\mathbb{Z}$ and the rationals $\mathbb{Q}$ are totally disconnected.

11. Prove Proposition 11.7.

12. Do you think that there are any uncountable, closed, totally disconnected subsets of $\mathbb{R}$? (Think about this question before discovering the answer in the next chapter.)

Chapter 12

The Cantor Set and Fractals

12.1. The Cantor set. You are now ready to see the Cantor set C, the prototypical fractal and one of the most interesting and amazing sets in analysis. Here's how you construct it. Start with the closed unit interval as in Figure 12.1. Remove the open middle third $(1/3, 2/3)$, leaving two closed intervals of length $1/3$. Remove the open middle third of each, leaving four closed intervals of length $1/9$. Continue. At the nth step you have a set S_n consisting of 2^n closed intervals of length $1/3^n$. Let $C = \bigcap S_n$.

As a subset of the unit interval, C is bounded. As an intersection of closed sets, C is closed. Hence C is compact.

C contains countably infinitely many boundary points of intervals, such as 0, 1, 1/3, 2/3, 1/9, 2/9, and so on. However, C contains lots more points:

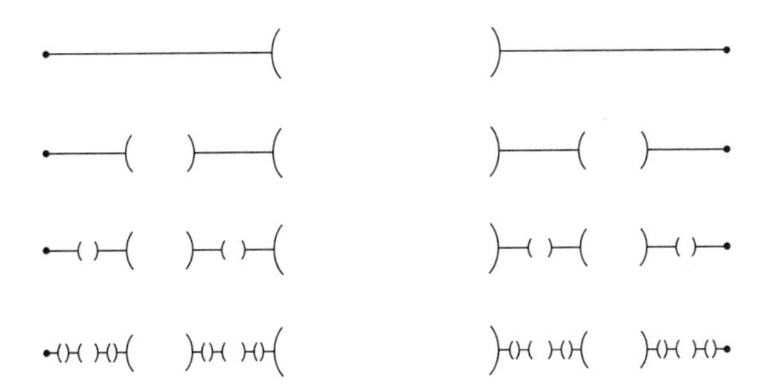

Figure 12.1. The Cantor set C comes from the unit interval by successively removing middle thirds forever.

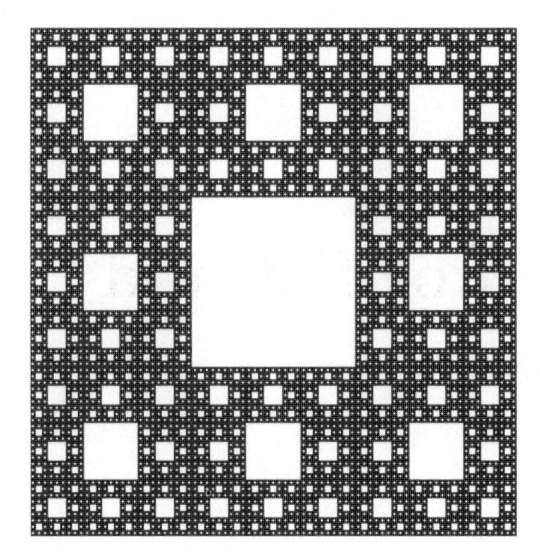

Figure 12.2. Sierpinski's Carpet has dimension $\log_3 8 \approx 1.9$. (From `http://en.wikipedia.org/wiki/Sierpinski_carpet`.)

12.2. Proposition. *The Cantor set is uncountable.*

Proof. Represent elements of the Cantor set as base 3 decimals such as $.021201222\ldots$. At the first step in the construction of the Cantor set we removed decimals with a 1 in the first decimal place. At the second step we removed remaining decimals with a 1 in the second decimal place. Hence C consists of base 3 decimals consisting just of 0s and 2s, such as $.020020222\ldots$.

The rest of the proof is like the proof of Proposition 2.4 that the set $\mathbb{R}$ of reals is uncountable. Suppose that C is countable, listed. Construct an element of C not on the list by making the first digit different from the first one on the list, the second digit different from the second one on the list, and so on. Therefore C is not countable. □

Although uncountable, the Cantor set has length or measure 0. To see this, just note that S_1 has measure $2/3$, that S_2 has measure $(2/3)^2$, and S_n has measure $(2/3)^n$, which approaches 0. Since C is contained in each S_n, its measure must be 0.

(*Measure* is a general word that applies in all dimensions, meaning length or area or volume. A rigorous treatment requires substantial mathematical development, as in a course on "measure theory.")

We also note that *the Cantor set is totally disconnected* because between any two elements there are deleted intervals.

Figure 12.3. The Menger sponge has dimension $\log_3 20 \approx 2.7$. © Paul Bourke. `http://astronomy.swin.edu.au/~pbourke/index.html`

12.3. Dimension and fractals (informal discussion). The plane is 2-dimensional, and a region in the plane is 2-dimensional. A curve in the plane, however, is just 1-dimensional. Likewise, a line or the interval $[0, 1]$ is 1-dimensional. A single point is 0-dimensional. (Some people say that the empty set has dimension -1, whatever that means.) What about the Cantor set? It seems higher dimensional than a point, but lower dimensional than the interval $[0, 1]$. In fact, it has dimension $\log_3 2 \approx .63$, according to an advanced mathematical concept of "Hausdorff dimension." It is the prototypical *fractal*, or fractional dimensional set. Like many fractals, it has self similarity: it is made up of two smaller copies of itself, each $1/3$ as large as the whole. There is a nice formula for the dimension of such a self-similar set:

12.4. Theorem. *A self-similar set made up of p smaller copies of itself, each $1/q$ as large as the whole, has dimension $\log_q p$.*

Figure 12.4. A random fractal landscape by R. F. Voss from *The Fractal Geometry of Nature* by Benoit Mandelbrot. Used by permission.

That the Cantor set has dimension $\log_3 2 \approx .63$ is one example. The unit interval is made up of n smaller copies, each $1/n$ as large, for dimension $\log_n n = 1$. A unit square region is made up of n^2 smaller copies, each $1/n$ as large, for dimension $\log_n n^2 = 2$.

The proof of Theorem 12.4 is beyond the scope of this text.

12.5. More fractals. Figure 12.2 shows a planar fractal called Sierpinski's carpet. It is made up of 8 smaller copies of itself, each 1/3 as large. Hence by Theorem 12.4 its dimension is $\log_3 8 \approx 1.9$.

Figure 12.3 shows a spatial fractal called the Menger sponge. It is made up of 20 smaller copies of itself, each 1/3 the size. Hence its dimension is $\log_3 20 \approx 2.7$.

Benoit Mandelbrot, the Father of Fractals, discovered that fractals provide better models of physical reality than the smooth surfaces of calculus. Figure 12.4 from his famous book on *The Fractal Geometry of Nature* shows a random fractal mountain which looks much more realistic than the upside-down paraboloids of calculus books.

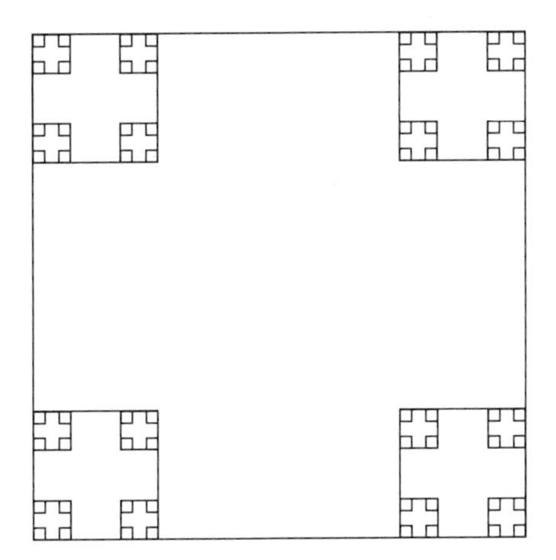

Figure 12.5. A fractal constructed by repeatedly removing everything but four corners each $1/4$ as large. (From Morgan's *Geometric Measure Theory* book.)

Exercises 12

1. Show that every point of the Cantor set is an accumulation point. (Such closed sets are called perfect. Every nonempty perfect real set is uncountable.)

2. Construct a version of the Cantor set by removing middle fifths instead of middle thirds. Is it still compact, uncountable, totally disconnected, and of measure 0?

3. Construct a version of the Cantor set by starting with $[0, 1]$, removing a middle interval of length $1/4$, then removing two middle intervals of total length $1/8$, then removing four middle intervals of total length $1/16$. Is it still compact, uncountable, and totally disconnected? What is its measure?

4. What is the dimension of the fractal of Figure 12.5?

5. Give an example of a totally disconnected set $S \subset [0, 1]$ whose closure is the whole interval.

6. Give examples, if possible, of continuous maps of the Cantor set onto $\mathbb{R}$, $(0, 1)$, and $[0, 1]$, or say why not possible.

Part III

Calculus

Part III: Calculus. It is satisfying to see how naturally the theory of calculus follows from the basic concepts of real analysis, especially when you remember that after Newton and Leibniz invented the calculus in the late 1600s, it took mathematicians two hundred years to get the theory straight.

This part focuses on $\mathbb{R}^1$ rather than $\mathbb{R}^n$, although Chapters 16 and 17 hold for real-valued functions on $\mathbb{R}^n$ as well.

Chapter 13

The Derivative and the Mean Value Theorem

Differentiation and integration are the two distinguishing processes of calculus. The first major theorem about the derivative, the Mean Value Theorem, follows easily from the compactness of the interval $[a, b]$, via Proposition 13.2 and Rolle's Theorem 13.3. We begin by recalling the definition of the derivative of a function on an interval in $\mathbb{R}$.

13.1. Definition. Let f be a real-valued function on an open interval (a, b). Define the *derivative* $f' = df/dx$ by

$$f'(x) = \lim_{t \to x} \frac{f(t) - f(x)}{t - x} = \lim_{\Delta x \to 0} \frac{\Delta f}{\Delta x}.$$

If this limit exists, we say that f is *differentiable* at x. One important interpretation of the derivative is the slope of the graph. More generally, the derivative gives a rate of change.

We assume the easy and familiar properties of the derivative and now state and prove the important theorems. We mention that *if f is differentiable at x then f is continuous at x.* If f has a continuous derivative, we say that f is *continuously differentiable* or C^1. If f has k continuous derivatives, we say that f is C^k. If f has derivatives of all orders, we say that f is C^∞ or *smooth.* (Sometimes *smooth* is used for C^1. It is a somewhat vague term that means one does not want to worry about differentiability and wants to assume whatever is needed.)

We say that f is *piecewise C^k* or smooth if the domain is the union of finitely many closed intervals on which f is C^k or smooth. (For this definition, at endpoints of a closed interval, we only consider how f changes as we move inside the interval.)

13.2. Proposition. *If f is differentiable at a local interior minimum (or maximum) point x, then $f'(x) = 0$.*

Proof. For t near a local minimum point x, the numerator in the definition of the derivative is nonnegative. For $t > x$, the denominator is positive, and hence $f'(x)$ is a limit of nonnegative numbers. For $t < x$, the denominator is negative, and hence $f'(x)$ is a limit of nonpositive numbers. The only possibility is $f'(x) = 0$. The proof is similar for a local maximum. □

Remark. Proposition 13.2 is the basis for finding maxima and minima in calculus. You just have to check where the derivative is 0 (or does not exist) for interior extrema, and also the endpoints (or extreme cases such as $x \to \pm\infty$).

13.3. Rolle's Theorem. *Suppose that f is continuous on $[a, b]$ and differentiable on (a, b), and that $f(a) = f(b)$. Then for some $c \in (a, b)$, $f'(c) = 0$.*

Proof. As a continuous function on a compact set, f has a maximum and a minimum. If the maximum occurs on the interior, by Proposition 13.2 f' vanishes (is zero) there. Likewise if the minimum occurs on the interior, f' vanishes there. Otherwise, the maximum and minimum both occur at the endpoints, f is constant, and f' vanishes everywhere. □

13.4. The Mean Value Theorem. *Suppose that f is continuous on $[a, b]$ and differentiable on (a, b). Then for some $c \in (a, b)$,*

$$f'(c) = \frac{f(b) - f(a)}{b - a}.$$

Proof. See Figure 13.1. By horizontal scaling and translation, we may assume that $[a, b] = [0, 1]$. Unless $f(0) = f(1)$ (in which case the result follows immediately from Rolle's Theorem), by vertical scaling and translation we may assume that $f(0) = 0$ and $f(1) = 1$. Let $g(x) = f(x) - x$. Then $g(0) = g(1) = 0$, so by Rolle's Theorem, for some $c \in (a, b)$, $0 = g'(c) = f'(c) - 1$. Therefore,

$$f'(c) = 1 = \frac{f(b) - f(a)}{b - a}.$$ □

The following theorem is the only reason you need the Mean Value Theorem to do calculus.

13.5. Corollary of the Mean Value Theorem. *On an interval where f' is always 0, f is constant.*

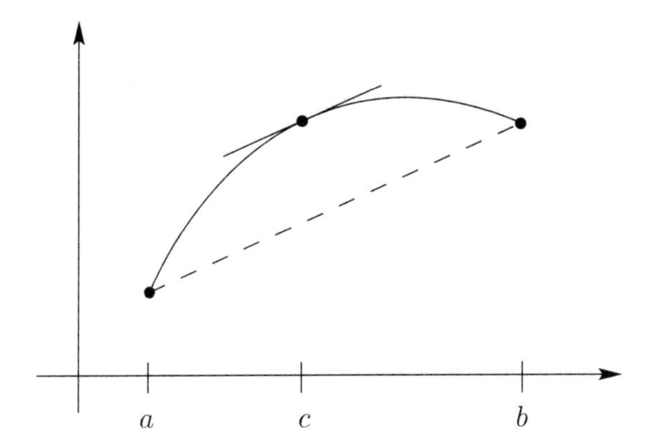

Figure 13.1. The Mean Value Theorem says that there is a point c where the instantaneous slope equals the average slope from a to b.

Proof. Suppose $a < b$ are any two points in the interval. By the Mean Value Theorem, $f(b) = f(a)$. □

13.6. The Cantor Function. Corollary 13.5 may sound obvious, but it is almost false. There is a nonconstant continuous function on $[0, 1]$ with derivative 0 except at a set of points of measure 0. The set is the Cantor set and the function is called the Cantor function. It is defined as follows (see Figure 13.2). Define f to be 0 at 0 and 1 at 1. On the middle third, define f to be $1/2$. On the middle thirds of the remaining two intervals, define f to be $1/4$ and $3/4$. On the middle thirds of the remaining four intervals, define f to be $1/8$, $3/8$, $5/8$, and $7/8$. Continue. f extends to a continuous function on $[0, 1]$. On the (open) complement of the Cantor set, f is constant on intervals, and hence has derivative 0.

The function $g(x) = x/2 - f(x)$ is the one secretly referred to in the Preface. Its derivative is $1/2$ (except in the measure-0 Cantor set), but it decreases from $g(0) = 0$ to $g(1) = -1/2$.

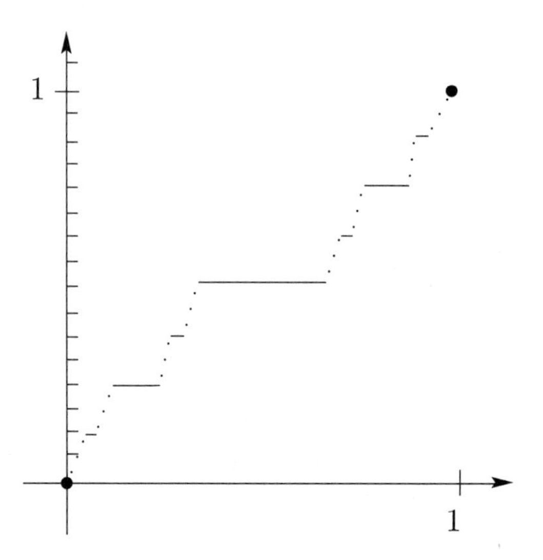

Figure 13.2. The Cantor function f has derivative 0 except on the Cantor set of measure 0, but it is not constant.

Exercises 13

1. Check the Mean Value Theorem for the function $f(x) = x^3$ on $[0,1]$.

2. Suppose that $f\colon \mathbb{R} \to \mathbb{R}$ satisfies $f(0) = 0$ and $|f'(x)| \leq M$. Prove that $|f(x)| \leq M|x|$. Apply this to the function $f(x) = \sin x$.

3. Discuss the logical chain of reasoning from compactness of $[a,b]$ to Corollary 13.5.

4. Show that the Cantor function is a continuous map of the Cantor set onto $[0,1]$, solving part of Exercise 12.6.

5. Define a map f from C into $[0,1]$ as follows. Given a point in the Cantor set, represent it by a base three decimal without 1s in it, such as $.0222022002\ldots$, change the 2s to 1s to get something like .0111011001, and then interpret it as a base two decimal in [0,1]. Is f continuous? surjective? Is f related to the Cantor function?

Chapter 14

The Riemann Integral

This chapter defines the standard, Riemann integral of a function on a bounded interval $[a, b]$ in $\mathbb{R}$ and shows that the process works for every continuous function f on $[a, b]$, using the fact that a continuous function on a compact set is uniformly continuous.

14.1. The Riemann integral. The Riemann integral $\int_a^b f(x)\,dx$ of a function f over an interval $[a, b]$ represents the area under the graph. The area may be approximated as in Figure 14.1 by chopping the interval up into narrow subintervals of perhaps variable thickness Δx, approximating each subarea by a skinny rectangle of height $f(x)$, thickness Δx, and area $\Delta A = f(x)\Delta x$, and adding them up:

$$A \approx \sum f(x)\Delta x.$$

This approximating sum is called the *Riemann sum.* To get the exact area, we take the limit as the maximum thickness goes to 0, and call this the *Riemann integral:*

$$\int_a^b f(x)\,dx = \lim_{\Delta x \to 0} \sum f(x)\Delta x.$$

The limit must be independent of the choice of subintervals and of the choice of where we evaluate $f(x)$ in each subinterval. If the limit exists, we say that f is *integrable* on $[a, b]$.

If $f(x)$ is a constant c, then f is integrable and

$$\int_a^b f(x)\,dx = c(b - a).$$

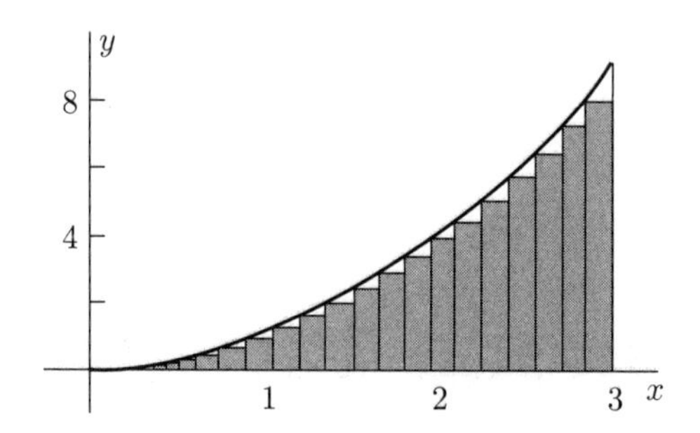

Figure 14.1. The area under the graph of f may be approximated by the sum of areas of skinny rectangles of height $f(x)$ and thickness Δx.

Indeed, every Riemann sum

$$\sum f(x)\Delta x = c\sum \Delta x = c(b-a).$$

This corresponds to the fact that the area of a rectangle of height c and width $b-a$ is $c(b-a)$.

14.2. Nonintegrable functions. One function which is not integrable on $[0,1]$ is the characteristic function $\chi_{\mathbb{Q}}$ of the rationals, which equals 1 on the rationals and 0 off the rationals. Indeed, no matter how small Δx, there are Riemann sums equal to one, obtained by always choosing to evaluate $\chi_{\mathbb{Q}}(x)$ at a rational, and there are Riemann sums equal to zero, obtained by always choosing to evaluate $\chi_{\mathbb{Q}}(x)$ at an irrational.

Another function which is not integrable on $[0,1]$ is the function $f(x) = 1/\sqrt{x}$ because no matter how small Δx, there are Riemann sums arbitrarily large, obtained by choosing to evaluate $f(x)$ at a point very close to 0 on the first subinterval. In general, *an integrable function must be bounded* (Exercise 3).

Fortunately, every continuous function is integrable:

14.3. Theorem. *Every continuous function is integrable on* $[a,b]$.

Proof. By Exercise 4.6, it suffices to show that any sequence of Riemann sums with $\Delta x \to 0$ is Cauchy. By Theorem 10.2, the continuous function f on the compact set $[a,b]$ is uniformly continuous: given $\varepsilon > 0$, there exists $\delta > 0$ such that

$$|\Delta x| < \delta \Rightarrow |\Delta f| < \varepsilon. \tag{1}$$

Now consider two Riemann sums both with $|\Delta x| < \delta/2$. Their subintervals intersect in smaller subintervals of thickness $|\Delta x| < \delta/2$. On each such subinterval, the values $f(x)$ from the two Riemann sums come from points at distance at most $\delta/2$ from a point in the intersection and hence at distance at most δ from each other. By (1), these values differ by at most ε. Summing over the smaller subintervals, we see that the absolute value of the difference of the Riemann sums equals

$$\left|\sum \Delta f \, \Delta x\right| \leq \sum |\Delta f| \, \Delta x \leq \sum \varepsilon \, \Delta x = \varepsilon \sum \Delta x = \varepsilon(b-a).$$

Therefore the sequence is Cauchy as desired and we conclude that the function is integrable. □

Remark. The whole truth is that a function on $[a, b]$ is integrable if and only if it is bounded and its set of discontinuities has measure 0.

14.4. Negative functions. Although for the motivating interpretation as area it is easier to think of f as positive, everything applies as well to functions allowed negative values. Of course if the function is negative, the integral will be negative. Some students like to remember this as, "Area below the x-axis counts negative."

Normally we expect that $a < b$, but it is convenient to define the Riemann integral for $a > b$ by

$$\int_a^b f(x)\,dx = -\int_b^a f(x)\,dx.$$

Finally, instead of $\int_a^b f(x)\,dx$, we sometimes just write $\int_a^b f$.

14.5. Proposition. *Suppose that f and g are integrable on the relevant intervals and that C is a constant.*

a. $\int_a^b Cf = C\int_a^b f$.

b. $\int_a^b f + g = \int_a^b f + \int_a^b g$.

c. $\int_a^c f = \int_a^b f + \int_b^c f$.

d. $\left|\int_a^b f\right| \leq \int_a^b |f|$.

e. *If $f \leq g$ and $a < b$, then $\int_a^b f \leq \int_a^b g$.*

Proof. Most of these facts follow immediately from the corresponding facts about Riemann sums. For a, the Riemann sums for Cf are just C times the Riemann sums for f. For b, the Riemann sums for $f + g$ are sums of Riemann sums for f and for g. For c, Riemann sums for the second and third terms yield Riemann sums for the first. For d, the inequality holds for Riemann sums because the absolute value of a sum is less than or equal to the sums of the absolute values. For e, the inequality holds for Riemann sums with the same subintervals and the same choice of places to evaluate $f(x)$ and $g(x)$. □

14.6. Corollary. *Suppose that f is continuous on $[a, b]$. Then*

$$(b-a) \min_{a \leq x \leq b} f(x) \leq \int_a^b f(x)\, dx \leq (b-a) \max_{a \leq x \leq b} f(x).$$

Proof. By Proposition 14.5e, taking g to be the constant $\max_{a \leq x \leq b} f(x) = c$,

$$\int_a^b f(x)\, dx \leq \int_a^b c\, dx = c(b-a),$$

proving the second inequality. The proof of the first is similar. □

Exercises 14

1. In approximating $\int_0^4 x\,dx$, suppose you divide $[0,4]$ into four unit subintervals. What are the possible values of Riemann sums?

2. Compute directly from the definition that

$$\int_a^b c\,dx = c(b-a).$$

3. Compute directly from the definition that

$$\int_0^1 x^2\,dx = 1/3$$

by the following steps. Since by Theorem 14.3 the integral exists independent of choice of subintervals or points to evaluate $f(x)$, we may choose to divide $[0,1]$ into n subintervals of length $\Delta x = 1/n$ and evaluate at the right-hand endpoints.

a. Show that the Riemann sums equal

$$\sum f(x)\Delta x = \sum_{k=1}^{n} (k/n)^2(1/n) = \frac{1}{n^3}\sum_{k=1}^{n} k^2.$$

b. Use the formula $\sum_{k=1}^{n} k^2 = \frac{n(n+1/2)(n+1)}{3}$ and take the limit as $n \to \infty$ to conclude that $\int_0^1 x^2\,dx = 1/3$.

4. Prove that a nonnegative Riemann integrable function is bounded.

5. Prove that every Riemann integrable function is bounded.

6. Explain why the definition of the Riemann integral does not apply to unbounded intervals such as $[0,\infty)$. (Sometimes such an "improper" integral is defined as $\lim_{R\to\infty} \int_0^R f(x)\,dx$.)

7. Is the function $f\colon [0,1] \to \mathbb{R}$ defined by

$$f(x) = \begin{cases} 0 & \text{for } 0 \le x \le 1/2, \\ 1 & \text{for } 1/2 < x \le 1 \end{cases}$$

integrable?

Chapter 15

The Fundamental Theorem of Calculus

The Fundamental Theorem of Calculus is most popular for its second part, which says that you can integrate just by antidifferentiating, instead of doing painful limits of Riemann sums. Both parts essentially say that integration and differentiation are opposites. It is quite remarkable that there is any relationship between integration and differentiation, between area and slope, between limits of Riemann sums and limits of ratios of change.

For the first part, to be able to differentiate after integrating, we treat the upper limit of integration b as a variable, and think of how the Riemann integral changes as b changes.

15.1. The Fundamental Theorem of Calculus. *Let f be a continuous function on an open interval containing $[a, b]$.*

I. $\frac{d}{db} \int_a^b f(x)\, dx = f(b)$.

II. *If $f(x) = F'(x)$, then $\int_a^b f(x)\, dx = F(x)|_a^b = F(b) - F(a)$.*

Proof of I. By the definition of the derivative,

$$\begin{aligned} \frac{d}{db} \int_a^b f(x)\, dx &= \lim_{\Delta b \to 0} \frac{\int_a^{b+\Delta b} f(x)\, dx - \int_a^b f(x)\, dx}{\Delta b} \\ &= \lim_{\Delta b \to 0} \frac{\int_b^{b+\Delta b} f(x)\, dx}{\Delta b}. \end{aligned}$$

If $\Delta b > 0$, by Corollary 14.6,

$$\min_{|x-b|\leq|\Delta b|} f(x) \leq \frac{\int_b^{b+\Delta b} f(x)\,dx}{\Delta b} \leq \max_{|x-b|\leq|\Delta b|} f(x).$$

If $\Delta b < 0$, the sign of both the numerator and the denominator change, the fraction remains unchanged, and the inequalities still hold. As $\Delta b \to 0$, the left-hand side and the right-hand approach $f(b)$; hence so does the fraction. Therefore

$$\frac{d}{db}\int_a^b f(x)\,dx = \lim_{\Delta b\to 0}\frac{\int_b^{b+\Delta b} f(x)\,dx}{\Delta b} = f(b)$$

as desired. □

Proof of II. Note that by I,

$$\frac{d}{db}\left(F(b) - \int_a^b f(x)\,dx\right) = F'(b) - f(b) = f(b) - f(b) = 0.$$

By Corollary 13.5 to the Mean Value Theorem,

$$F(b) - \int_a^b f(x)\,dx = C.$$

Plugging in $b = a$ yields

$$F(a) = C.$$

Therefore

$$\int_a^b f(x)\,dx = F(b) - F(a).$$ □

15.2. Remark. Why don't we just *define* $\int_a^b f(x)\,dx$ as $F(b)-F(a)$? First of all, we may not know of any antiderivative F. Second of all, how could we expect *that* to have anything to do with area or other applications? The only sound approach is to define area somehow (the Riemann integral) and then figure out an easy way to compute it.

15.3. Remark. There are amazing generalizations of the Fundamental Theorem to functions on domains in $\mathbb{R}^n$ and more general domains, which go by names such as Green's Theorem, Gauss's Theorem or the Divergence Theorem (29.5), and Stokes's Theorem.

Exercises 15

1. Use the Fundamental Theorem to compute that $\int_0^1 x^2\,dx = 1/3$. Is this easier than direct computation from the definition?

2. Compute $\frac{d}{db}\int_a^b x^2\,dx$ two ways, first using 15.1(II), then using 15.1(I).

3. Let $F(x) = \int_0^x e^{-t^2}\,dt$. Compute $F'(x)$ and $F'(0)$.

4. What is $\frac{d}{da}\int_a^b f(x)\,dx$?

Chapter 16

Sequences of Functions

What does it mean for a sequence of *functions* to converge to some limit function? There is an easy definition, just looking at one point at a time.

16.1. Definition (pointwise convergence). A sequence of functions f_n converges to f *pointwise* on some domain E if for every $x \in E$, the sequcnce of nnmbers $f_n(x)$ converges to $f(x)$; i.e., for every x in E, given $\varepsilon > 0$, there is an N, such that

$$n > N \Rightarrow |f_n(x) - f(x)| < \varepsilon.$$

For example, if $f_n(x) = x/n$, then $f_n \to f$, where $f(x) = 0$, on $\mathbb{R}$. As a second example, if $f_n(x) = x^n$, as in Figure 16.1, then on $[0, 1]$ $f_n \to f$, where

$$f(x) = \begin{cases} 0 & \text{for } 0 \leq x < 1 \\ 1 & \text{for } x = 1. \end{cases}$$

It is somewhat disturbing that this limit of continuous functions is not itself continuous. The problem is that although $f_n \to f$ pointwise, the closer the point x gets to 1, the longer it takes for $f_n(x)$ to converge to $f(x)$. Indeed, how big you have to take n to make $f_n(x)$ close to $f(x)$ *goes to infinity* as x approaches 1. We need a more uniform kind of convergence.

16.2. Definition (uniform convergence). A sequence of functions $f_n \to f$ *uniformly* on some domain E if given $\varepsilon > 0$, there is an N, such that

$$n > N \Rightarrow |f_n(x) - f(x)| < \varepsilon$$

for all x in E.

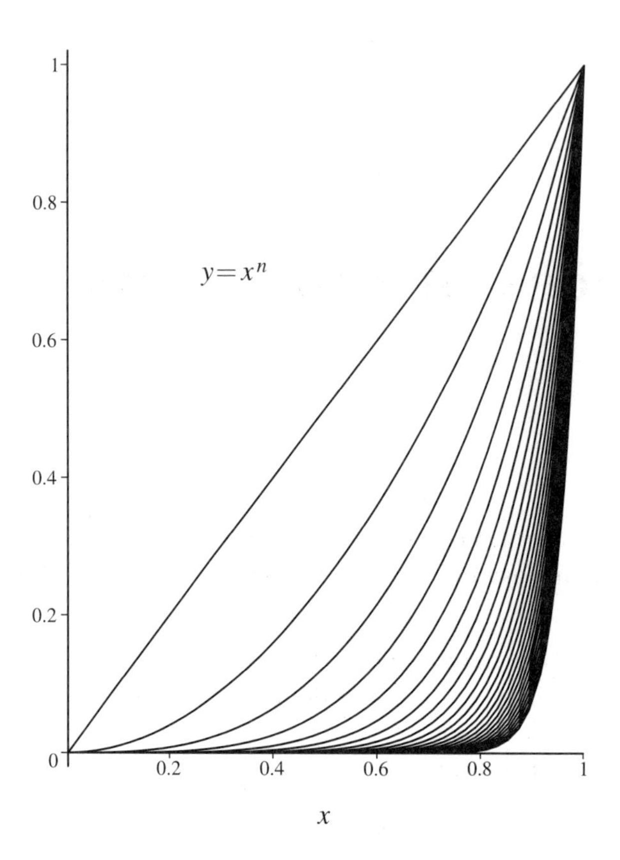

Figure 16.1. The continuous functions $f_n(x) = x^n$ converge pointwise to a discontinuous function.

The only difference in the two definitions is that in the second, N appears before x, and hence the same N must work for all x, whereas in the first definition, N is allowed to depend on x. It is amazing that such a seemingly small change can be so important. Uniform convergence is just what you need to get continuous limits:

16.3. Theorem. *A uniform limit of continuous functions is continuous.*

Proof. The idea of the proof is that by uniform convergence, we can handle all points near any particular point p by looking at one f_n that is uniformly near f.

Given $\varepsilon > 0$, choose N such that for all x in the domain D,

$$n > N \Rightarrow |f_n(x) - f(x)| < \varepsilon.$$

Since f_{N+1} is continuous, given a point p in D, we can choose $\delta > 0$ such that

$$|x - p| < \delta \Rightarrow |f_{N+1}(x) - f_{N+1}(p)| < \varepsilon.$$

Then if $|x - p| < \delta$,

$$\begin{aligned}|f(x) - f(p)| &\le |f(x) - f_{N+1}(x)| + |f_{N+1}(x) - f_{N+1}(p)| + |f_{N+1}(p) - f(p)| \\ &< \varepsilon + \varepsilon + \varepsilon = 3\varepsilon.\end{aligned}$$

□

Therefore f is continuous.

(It is important that by uniform convergence N does not depend on x because δ depends on N and δ must not depend on x.)

16.4. Question. Suppose a sequence of continuous functions f_n converges pointwise to a function f on $[0, 1]$. Is the limit of the integrals equal to the integral of the limit:

$$\lim \int_0^1 f_n = \int_0^1 \lim f \ ?$$

In other words, can we switch the limit and the integral?

Example. Consider our problematic $f_n(x) = x^n$, which converges pointwise but not uniformly to 0 on $(0, 1)$. Then

$$\lim \int_0^1 f_n = \lim \int_0^1 x^n \, dx = \lim \left[\frac{x^{n+1}}{n+1}\right]_0^1 = \lim \frac{1}{n+1} = 0,$$

while

$$\int_0^1 \lim f_n = \int_0^1 \lim 0 = 0.$$

Looks OK.

Example. On $[0, 1]$, let $f_n(x)$ be 0, except let $f_n(x)$ be n for $0 < x < 1/n$, as in Figure 16.2, so that the integral is always 1, while $f_n(x)$ converges pointwise to 0. Then

$$\lim \int_0^1 f_n = \lim 1 = 1,$$

while

$$\int_0^1 \lim f_n = \int_0^1 \lim 0 = 0.$$

So it does not always work. Fortunately, it does always work if the convergence is uniform:

16.5. Theorem. *For a uniform limit of continuous functions on a bounded interval $[a, b]$,*

$$\lim \int_a^b f_n = \int_a^b \lim f_n.$$

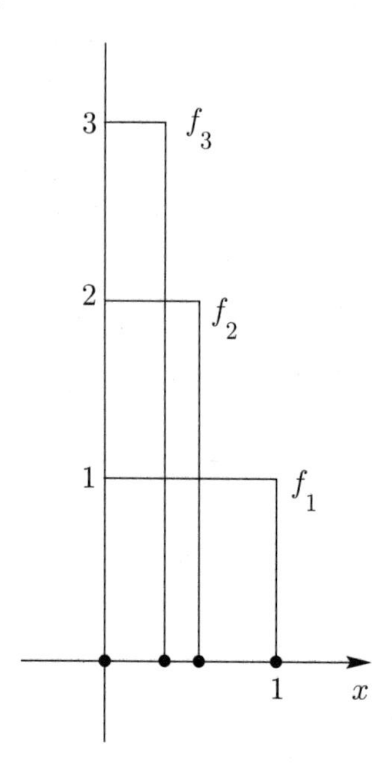

Figure 16.2. Functions $f_n(x)$ with integral 1 and limit 0.

Proof. Given $\varepsilon > 0$, choose N such that if $n > N$, $|f_n - f| < \varepsilon$. Then

$$\left|\int f_n - \int f\right| = \left|\int (f_n - f)\right| \leq \int |f_n - f| < \int \varepsilon = \varepsilon(b-a). \qquad \square$$

Example.

$$\lim_{n\to\infty} \int_0^1 (x^2 + e^{x^2}/n)\, dx = \int_0^1 x^2\, dx = 1/3$$

because $x^2 + e^{x^2}/n \to x^2$ uniformly. Indeed,

$$|(x^2 + e^{x^2}/n) - x^2| = |e^{x^2}/n| \leq e/n,$$

which gets small at a rate independent of x.

16.6. Remark. Even for uniform convergence, the *derivative* of the limit need not equal the limit of the derivatives (Exercise 9), but something is still true, as we'll see for example in Theorem 20.7.

Exercises 16

1. Give an example of a sequence f_n of functions on $\mathbb{Z}$ which converge pointwise but not uniformly.

2. Give an example of a pointwise convergent sequence f_n of functions on $[0, 1]$ for which

$$\lim \int_0^1 f_n \neq \int_0^1 \lim f_n.$$

3. Consider the statement: If $f_n \to f$ uniformly, then $f_n^2 \to f^2$ uniformly. First give a counterexample. Second add a simple hypothesis and prove the revised statement.

4. Prove or give a counterexample: If nonvanishing (never 0) functions $f_n \to f$ uniformly, then $1/f_n \to 1/f$ uniformly.

5. Consider continuous functions from $[0, 1]$ to $\mathbb{R}$

$$0 \leq f_1 \leq f_2 \leq f_3 \leq \cdots$$

converging pointwise to f. Must f be continuous? Give a proof or a counterexample.

6. Find $\lim_{n\to\infty} \int_0^1 \frac{e^{x^4}}{n}\, dx$. Justify.

7. Find $\lim_{n\to\infty} \int_1^2 x^{2-(\sin nx)/n}\, dx$. Justify.

8. A function is called *Lipschitz* with Lipschitz constant C if

$$|f(x) - f(y)| \leq C|x - y|$$

for all x, y in the domain. Let f_n be a pointwise convergent sequence of Lipschitz functions on $[0, 1]$ with Lipschitz constant C. Prove that the sequence converges uniformly.

9. Show by counterexample that even if differentiable functions f_n converge uniformly to a differentiable function f on $\mathbb{R}$, it does not follow that the derivatives f_n' converge uniformly to f'. In particular, describe such a sequence of functions f_n converging to 0 such that $f_n'(0) = 1$ for every n.

Chapter 17

The Lebesgue Theory

The Riemann integral, however natural, has certain technical flaws and complications alleviated by the more general Lebesgue theory. Although the Lebesgue theory is properly the subject of a graduate course, we present some of the convenient results here without proof so that you can start taking advantage of them right away.

The Lebesgue Integral is a generalization of the Riemann integral. For any Riemann integrable function, it gives the same answer. However, there are more integrable functions. The Lebesgue integral can ignore a countable set (or any set of length or measure 0). For example, in the Lebesgue theory, $\int_0^1 \chi_{\mathbb{Q}} = 0$, as it should, since the rationals are a negligible, countable set. Similarly, for the Cantor set C, $\int_0^1 \chi_C = 0$. We also allow unbounded functions and unbounded intervals, so that for example

$$\int_0^1 x^{-1/2} = 2x^{1/2}|_0^1 = 2,$$

and

$$\int_1^\infty x^{-2} = -x^{-1}|_1^\infty = 1.$$

For switching the limit and the integral, there is a stronger theorem with hypotheses that are easier to check:

17.1. Lebesgue's Dominated Convergence Theorem. *Let f_n be functions on a domain in $\mathbb{R}$ converging pointwise to a limit f. If there is a function g with finite integral such that each $|f_n| \le g$, then*

$$\int \lim f_n = \lim \int f_n.$$

17.2. Corollary (Uniformly bounded functions on a bounded interval). *Let f_n be functions on $[a,b]$ converging pointwise to a limit f. Suppose that for some $M > 0$, every $|f_n| \leq M$. Then*

$$\int \lim f_n = \lim \int f_n.$$

Proof. Apply Lebesgue's Dominated Convergence Theorem, with $g(x) = M$. Since the domain is bounded, M has finite integral. □

For switching the order of integration in a double integral, the following theorem is very useful.

17.3. Fubini's Theorem. *For a double integral, it is OK to switch the order of integration if either order yields a finite answer when the integrand is replaced by its absolute value.*

17.4. Caveat on measurability. Technically, Theorems 17.1–17.3 have an additional hypothesis, that the integrands be "measurable." Measurable functions include continuous functions, piecewise continuous functions, perhaps altered on countable sets or sets of measure 0, and much more. They include all functions that you have ever heard of and all functions that can arise in the real world. Indeed, there is a theory of mathematics (without the Axiom of Choice) in which all functions are measurable. So this is not something to worry about.

We give one more very useful theorem for switching integration with respect to one variable x with differentiation with respect to another variable t.

17.5. Leibniz's Rule. Suppose that $\int_{a(t)}^{b(t)} f(x,t)\,dx$ exists, that $a(t)$, $b(t)$, and $f(x,t)$ are all continuously differentiable with respect to t, and that there is a function $g(x) \geq \left|\frac{\partial f}{\partial t}(x,t)\right|$ with $\int_{a(t)}^{b(t)} g(x)\,dx < \infty$. Then

$$\frac{d}{dt}\int_{a(t)}^{b(t)} f(x,t)\,dx = \int_{x=a(t)}^{b(t)} \frac{\partial f}{\partial t}(x,t)\,dx + f(b(t),t)\,b'(t) - f(a(t),t)\,a'(t).$$

We allow $a = -\infty$ or $b = +\infty$.

Exercises 17

1. Compute $\lim_{n\to\infty} \int_1^2 x^{2-(\sin nx)/n}\,dx$. Justify.

2. Compute $\lim_{n\to\infty} \int_1^\infty \frac{1+(-1)^n e^{-nx^2}}{x^2}\,dx$. Justify.

3. Compute $\lim_{n\to\infty} \int_0^1 \frac{1}{nx}\,dx$. Justify.

4. Compute $\int_{y=0}^{10} \int_{x=0}^{\pi/3} xy\cos xy^2\,dx\,dy$. Justify.
Answer: $3/400$.

5. Compute $\int_{y=0}^{1} \int_{x=y}^{1} \frac{x^2}{y^2} e^{-x^2/y}\,dx\,dy$. Justify.
Answer: $1 - 1/e$.

6. Compute $\frac{d}{dt} \int_{x=1}^{2} \frac{\sin x^2 t}{x}\,dx$. Justify.
Answer: $\frac{1}{2t}(\sin 4t - \sin t)$ for $t \neq 0$, $\frac{3}{2}$ for $t = 0$.

7. Compute $\frac{d}{dt} \int_{x=1}^{2} \frac{1}{x} e^{-x^2 t}\,dx$ at $t = 1$. Justify.
Answer: $-\frac{e^3-1}{2e^4} \approx -.17478$.

8. Compute $\frac{d}{dy} \int_{x=2+y}^{\infty} \frac{dx}{x^2+y^2}$ at $y = 0$. Justify.
Answer: $-1/4$.

9. Give a counterexample to the following converse to Lebesgue's Dominated Convergence Theorem: If $f_n \to f$ and $\int f_n \to \int f$ (all finite), then there exists $g \geq |f_n|$ with $\int g < \infty$.

Chapter 18

Infinite Series $\sum_{n=1}^{\infty} a_n$

Infinite sums or series turn out to be very useful and important because in the limit you can actually attain an apparently unattainable value or function by adding on tiny corrections forever. In theory, they are no harder than sequences, because to see if a series converges you just look at the sequence of subtotals or partial sums. For example, we say that the infinite series

$$\frac{1}{2} + \frac{1}{4} + \frac{1}{8} + \frac{1}{16} + \cdots$$

converges to 1 because the sequence of partial sums

$$\frac{1}{2}, \frac{3}{4}, \frac{7}{8}, \frac{15}{16}, \ldots$$

converges to 1.

18.1. Definition. The series $\sum_{n=1}^{\infty} a_n$ *converges to* L if given $\varepsilon > 0$ there is an N_0 such that

$$N > N_0 \Rightarrow \left| \sum_{n=1}^{N} a_n - L \right| < \varepsilon.$$

Otherwise the series *diverges* (perhaps to $+\infty$, perhaps to $-\infty$, or otherwise "by oscillation").

Of course if a series converges, the terms must approach 0 (Exercise 12). The converse fails. Just because the terms approach 0, the series need not converge. For example, see the harmonic series below.

18.2. Proposition. *Suppose that* $\sum_{n=1}^{\infty} a_n$ *converges to* A *and* $\sum_{n=1}^{\infty} b_n$ *converges to* B. *Then* $\sum_{n=1}^{\infty} ca_n$ *converges to* cA *and* $\sum_{n=1}^{\infty} a_n + b_n$ *converges to* $A + B$.

Proof. This proposition follows immediately from similar facts about sequences, as you'll show in Exercise 12. □

We now summarize some of the famous convergence tests from calculus.

18.3. Geometric series. *A geometric series* (obtained by repeatedly multiplying the initial term a_0 by a constant ratio r)

$$\sum a_0 r^n = a_0 + a_0 r + a_0 r^2 + a_0 r^3 + \cdots, \tag{1}$$

converges to $\frac{a_0}{1-r}$ *if* $|r| < 1$ *and diverges if* $|r| \geq 1$ (*and* $a_0 \neq 0$). For example, the series at the beginning of Chapter 18, a geometric series with initial term $a_0 = 1/2$ and ratio $r = 1/2$, converges to $\frac{1/2}{1-1/2} = 1$.

Proof. If $|r| \geq 1$ (and $a_0 \neq 0$), the terms do not approach 0, so the series diverges. Suppose $|r| < 1$. Let S_n denote the subtotal

$$\begin{aligned} S_n &= a_0 + a_0 r + a_0 r^2 + a_0 r^3 + \cdots + a_0 r^n; \\ r S_n &= \phantom{a_0 +{}} a_0 r + a_0 r^2 + a_0 r^3 + \cdots + a_0 r^n + a_0 r^{n+1}; \\ (1-r) S_n &= a_0 \qquad\qquad\qquad\qquad\qquad\qquad\qquad - a_0 r^{n+1}; \\ S_n &= \frac{a_0(1-r^{n+1})}{1-r}. \end{aligned}$$

As $n \to \infty$, $S_n \to \frac{a_0}{1-r}$. □

18.4. p-test. *The p-series* $\sum \frac{1}{n^p}$ *converges if* $p > 1$ *and diverges if* $p \leq 1$.

For example, taking $p = 1$, the **harmonic series**

$$\sum \frac{1}{n} = \frac{1}{1} + \frac{1}{2} + \frac{1}{3} + \frac{1}{4} + \cdots$$

diverges, despite the fact that the terms approach 0.

As a second example, taking $p = 2$,

$$\sum \frac{1}{n^2} = \frac{1}{1^2} + \frac{1}{2^2} + \frac{1}{3^2} + \frac{1}{4^2} + \cdots$$

converges, although we do not yet know what it converges to.

We omit the proof of the p-test, which uses comparison with integrals.

18.5. Comparison test. *If* $|a_n| \leq b_n$ *and* $\sum b_n$ *converges to* β, *then* $\sum a_n$ *converges, and its limit* $\alpha \leq \beta$.

For example, $\sum \frac{1}{1+n^2}$ converges by comparison with $\sum \frac{1}{n^2}$, which converges by the p-test.

Proof. It suffices to show that the sequence of subtotals, which is obviously bounded by β, is Cauchy. For $N_0 \leq M \leq N$,

$$\left|\sum_{n=1}^{N} a_n - \sum_{n=1}^{M} a_n\right| = \left|\sum_{n=M+1}^{N} a_n\right| \leq \sum_{n=M+1}^{N} |a_n| \leq \sum_{n=M+1}^{N} b_n = \left|\sum_{n=1}^{N} b_n - \sum_{n=1}^{M} b_n\right|,$$

which is small because the sequence of subtotals of $\sum b_n$ is Cauchy. □

Remark. The first million terms do not affect *whether* or not a series converges, although of course they do affect *what* it converges to. For example, $\sum_{n=2}^{\infty}\left(\frac{100}{\log_{10} n}\right)^n$ converges by comparison with $\sum_{n=2}^{\infty}\left(\frac{1}{2}\right)^n = \frac{1}{8}$ because for n large, $\frac{100}{\log_{10} n} < \frac{1}{2}$. The limit, however, is huge; just the single term when $n = 10^{10}$ is $\left(\frac{100}{10}\right)^{10^{10}} = 10^{10000000000}$.

18.6. Alternating series. *If the terms of a series $\sum a_n$ alternate in sign and $|a_n|$ decrease with limit* 0,

$$|a_1| \geq |a_2| \geq |a_3| \cdots \to 0,$$

then the series converges.

For example, the alternating harmonic series

$$\sum \frac{(-1)^{n+1}}{n} = \frac{1}{1} - \frac{1}{2} + \frac{1}{3} - \frac{1}{4} + \cdots$$

converges (to $\ln 2$).

Proof. We may assume that the first term is positive. The subtotals after an odd number of terms are decreasing and positive, and hence converge to a limit L_1. The subtotals after an even number of terms are increasing and bounded above by the first term, and hence converge to a limit L_2. Since their differences approach 0, $L_1 = L_2$ is the limit of the sequence of subtotals. □

Exercises 18

For Exercises 1–11, prove whether or not the series converges. If you can, give the limit.

1. $\frac{1}{10} + \frac{1}{100} + \frac{1}{1000} + \cdots$.

2. $3 + \frac{6}{5} + \frac{12}{25} + \frac{24}{125} + \cdots$.

3. $2 + 4 + 8 + 16 + \cdots$.

4. $\sum_{n=1}^{\infty} n^{1/n}$.

5. $\sum_{n=1}^{\infty} \frac{1}{n\sqrt{n}}$.

6. $\sum_{n=0}^{\infty} \frac{1}{n^3+2}$.

7. $\sum_{n=1}^{\infty} \frac{5 \ln n}{n^2}$.

8. $\sum_{n=2}^{\infty} \frac{(-1)^n}{\ln n}$. Why did we start the series with $n = 2$ instead of $n = 1$?

9. $\sum_{n=2}^{\infty} \frac{(-1)^n}{n^{1/n}}$.

10. $\sum_{n=3}^{\infty} \frac{\sin n}{(n^2+n-1)(\ln n)}$.

11. $\sum_{n=1}^{\infty} \frac{6n^2+89n+73}{n^4-213n}$.

12. Prove that if a series converges, the terms approach 0.

13. Prove Proposition 18.2.

Chapter 19

Absolute Convergence

19.1. Definitions. A series $\sum a_n$ *converges absolutely* if the series of absolute values $\sum |a_n|$ converges. Otherwise it *converges conditionally* (or diverges).

For example, the alternating harmonic series converges conditionally. It follows from the Comparison Test 18.5 that *absolute convergence implies convergence.*

A *rearrangement* of a series has exactly the same terms, but not necessarily in the same order. You might expect that order should not matter, and it does not if the series converges absolutely (Prop. 19.2), but it does matter if the series converges merely conditionally (Prop. 19.3). Here is an example:

$$\frac{1}{2}-\frac{1}{2}+\frac{1}{2}-\frac{1}{2}+\frac{1}{3}-\frac{1}{3}+\frac{1}{3}-\frac{1}{3}+\frac{1}{3}-\frac{1}{3}+\frac{1}{4}-\frac{1}{4}+\frac{1}{4}-\frac{1}{4}+\frac{1}{4}-\frac{1}{4}+\frac{1}{4}-\frac{1}{4}+\cdots$$

with partial sums

$$\frac{1}{2},\ 0,\ \frac{1}{2},\ 0,\ \frac{1}{3},\ 0,\ \frac{1}{3},\ 0,\ \frac{1}{3},\ 0,\ \frac{1}{4},\ 0,\ \frac{1}{4},\ 0,\ \frac{1}{4},\ 0,\ \frac{1}{4},\ 0,\ \ldots$$

converges to 0, but the rearrangement

$$\frac{1}{2}+\frac{1}{2}-\frac{1}{2}-\frac{1}{2}+\frac{1}{3}+\frac{1}{3}+\frac{1}{3}-\frac{1}{3}-\frac{1}{3}-\frac{1}{3}+\frac{1}{4}+\frac{1}{4}+\frac{1}{4}+\frac{1}{4}-\frac{1}{4}-\frac{1}{4}-\frac{1}{4}-\frac{1}{4}+\cdots$$

with partial sums

$$\frac{1}{2}, 1, \frac{1}{2}, 0, \frac{1}{3}, \frac{2}{3}, 1, \frac{2}{3}, \frac{1}{3}, 0, \frac{1}{4}, \frac{1}{2}, \frac{3}{4}, 1, \frac{3}{4}, \frac{1}{2}, \frac{1}{4}, 0, \ldots$$

diverges by oscillation.

19.2. Proposition. *If a series converges absolutely, then all rearrangements converge to the same limit.*

Proof. Let $\sum a_n$ denote the original series, which converges absolutely to some limit L, and let $\sum b_n$ denote the rearrangement. Given $\varepsilon > 0$, choose N_1 such that if $M, N > N_1$, then

$$\sum_{n=M}^{N} |a_n| < \varepsilon/2 \quad \text{and} \quad \left|\sum_{n=1}^{N} a_n - L\right| < \varepsilon/2. \tag{1}$$

Choose N_2 such that the first N_2 terms of $\sum b_n$ include the first N_1 terms of $\sum a_n$. Let $N > N_2$. Choose N_3 such that the first N_3 terms of $\sum a_n$ include the first N terms of $\sum b_n$. Then

$$\left|\sum_{n=1}^{N} b_n - L\right| \leq \left|\sum_{n=1}^{N} b_n - \sum_{n=1}^{N_3} a_n\right| + \left|\sum_{n=1}^{N_3} a_n - L\right|.$$

Since $N_3 > N_1$, by (1) the last expression is at most $\varepsilon/2$. Consider the middle expression. All of the terms of the first sum are included in the second sum, and the first N_1 terms of the second sum are included in the first sum. Hence by (1) the middle expression is at most $\varepsilon/2$. We conclude that

$$\left|\sum_{n=1}^{N} b_n - L\right| \leq \varepsilon,$$

which means that the rearrangement converges to the same limit L. □

19.3. Proposition. *Suppose that a series $\sum a_n$ converges conditionally. Then its terms may be rearranged to converge to any given limit, or to diverge to $\pm\infty$, or to diverge by oscillation.*

Proof sketch. First we claim that the amount of positive stuff, the sum of the positive terms, must diverge to $+\infty$, and that the amount of negative stuff must diverge to $-\infty$. If both converged, the series would converge absolutely. If just one converged, the series would diverge. Hence both must diverge.

Second, note that because the series converges, the terms approach 0.

Our rearrangements will keep the positive terms in the same order and the negative terms in the same order. To get big (positive) limits, we'll front load the positive terms. To get small (negative) limits, we'll front load the negative terms.

To get a rearrangement with prescribed limit L, take positive terms until you first pass L heading right. Then take negative terms until you first pass L heading left. Then take positive terms until you pass L heading right. Then take negative terms until you pass L heading left. Since there is an infinite amount of positive stuff and of negative stuff, you can continue forever. Since the terms go to 0, the amount by which you overshoot goes to 0. Hence the rearrangement converges to L.

To get a rearrangement which diverges to $+\infty$, take positive terms until you pass 1, then a negative term, then positive terms until you pass 2, then a negative term, then positive terms until you pass 3, and so on.

To get a rearrangement which diverges to $-\infty$, take negative terms until you pass -1, then a positive term, then negative terms until you pass -2, then a positive term, then negative terms until you pass -3, and so on.

To get a rearrangement which diverges by oscillation, take positive terms until you pass 1, then negative terms until you pass -1, then positive terms until you pass 1, then negative terms until you pass -1, and so on. □

19.4. Ratio test. *Given a series $\sum a_n$, let $\rho = \lim \left|\frac{a_{n+1}}{a_n}\right|$ (Greek letter rho). If $\rho < 1$, then the series converges absolutely. If $\rho > 1$, then the series diverges. If $\rho = 1$ or the limit does not exist, the test fails; the series could converge absolutely, converge conditionally, or diverge.*

For example, consider the series

$$\sum_{n=0}^{\infty} \frac{1}{n!} = \frac{1}{0!} + \frac{1}{1!} + \frac{1}{2!} + \frac{1}{3!} + \frac{1}{4!} + \cdots = \frac{1}{1} + \frac{1}{1} + \frac{1}{2} + \frac{1}{6} + \frac{1}{24} + \cdots.$$

To get to the nth term from the previous term, you multiply by $1/n$, so the limiting ratio $\rho = 0$. Consequently the series converges absolutely (to e).

Proof of Ratio test. Exercise 10. □

Although I don't usually teach the following test in calculus, it turns out to be of theoretical importance in real analysis.

19.5. Root test. *Given a series $\sum a_n$, let $\rho = \limsup \sqrt[n]{|a_n|}$. If $\rho < 1$, then the series converges absolutely. If $\rho > 1$, then the series diverges. If $\rho = 1$, the test fails; the series could converge absolutely, converge conditionally, or diverge.*

Proof. If $\rho < 1$, to give us some room to work, let $\rho < \sigma < 1$ (σ is the Greek letter sigma). Then for all large n, $\sqrt[n]{|a_n|} \le \sigma$, i.e., $|a_n| \le \sigma^n$, so that the series converges absolutely by comparison with the geometric series. If $\rho > 1$, for all large n, $|a_n| > 1$, and the series diverges because the terms do not approach 0. $\square$

Exercises 19

1. What are the possible values of rearrangements of the following series:

a. $\sum_{n=1}^{\infty} \frac{1}{2^n}$;

b. $\sum_{n=1}^{\infty} \frac{(-1)^n}{2^n}$;

c. $\sum_{n=1}^{\infty} \frac{(-1)^n}{\sqrt{n}}$;

d. $\sum_{n=1}^{\infty} \frac{1}{\sqrt{n}}$.

Prove whether the following series 2–9 converge absolutely, converge conditionally, or diverge. Give the limit if you can.

2. $\sum_{n=0}^{\infty} \frac{1}{5^n}$.

3. $\sum_{n=0}^{\infty} \frac{(-1)^n}{5^n}$.

4. $\sum_{n=0}^{\infty} \frac{5^n}{n!}$.

5. $\sum_{n=0}^{\infty} \frac{(-5)^n}{n!}$.

6. $\sum_{n=1}^{\infty} \frac{1}{n^n}$.

7. $\sum_{n=1}^{\infty} \frac{(-1)^n}{n^{1/5}}$.

8. $\sum_{n=1}^{\infty} \frac{(-1)^n}{1+1/n}$.

9. $\sum_{n=1}^{\infty} \frac{n \sin n}{e^n}$.

10. Prove the Ratio test 19.4.

Hint: Use the proof of the root test as a model.

Chapter 20

Power Series

Power series are an important kind of series of *functions* rather than just numbers. We begin with consideration of general series of functions.

20.1. Definitions. A series $\sum f_n$ of functions *converges* (pointwise or uniformly) if the sequence of partial sums converges (pointwise or uniformly). The series $\sum f_n$ *converges absolutely* if $\sum |f_n|$ converges absolutely. There is a nice test due to Weierstrass for uniform convergence, essentially by comparison with a series of constants.

20.2. Weierstrass M-test. *Suppose $|f_n| \leq M_n$ where each M_n is a positive number and $\sum M_n$ converges. Then $\sum f_n$ converges uniformly.*

Proof. For each x, $\sum f_n(x)$ converges by comparison with $\sum M_n$ to some $f(x)$. To prove the convergence uniform, note that

$$\left|\sum_{n=1}^{N} f_n(x) - f(x)\right| = \left|\sum_{n=1}^{N} f_n(x) - \sum_{n=1}^{\infty} f_n(x)\right| = \left|\sum_{n=N+1}^{\infty} f_n(x)\right| \leq \sum_{n=N+1}^{\infty} M_n,$$

which is small for N large, independent of x. □

For example, $\sum \frac{\sin nx}{n^2}$ converges uniformly on $\mathbb{R}$ by Weierstrass M-test comparison with $\sum \frac{1}{n^2}$. By Theorem 16.3, the limit is continuous.

20.3. Definition. A *power series* is a series $\sum a_n x^n$ of multiples of powers of x, such as

$$\sum_{n=0}^{\infty} \frac{1}{n!} x^n = 1 + x + \frac{1}{2}x^2 + \frac{1}{6}x^3 + \cdots,$$

which we will identify as e^x in Chapter 24. A function defined by a convergent power series is called *real analytic.*

Of course a power series is more likely to converge if x is small. You might expect a power series to converge absolutely for x small, converge conditionally for x medium, and diverge for x large. The truth is even a bit simpler.

20.4. Radius of convergence. *A power series has a* radius of convergence $0 \leq R \leq \infty$, *such that*

(1) *for* $|x| < R$, *the series converges absolutely;*

(2) *for* $|x| > R$, *the series diverges.*

For $x = \pm R$, *the series might converge absolutely, converge conditionally, or diverge.*

On every interval $[-R_0, R_0] \subset (-R, R)$, *the series converges uniformly and the series may be integrated term by term.*

Note that this result admits the possibilities that the series always converges (the case $R = \infty$) or never converges unless $x = 0$ (the case $R = 0$).

Proof. To prove (1) and (2), it suffices to show that if the series converges at x_0, then it converges absolutely for $|x| < |x_0|$. Since it converges at x_0, the terms $a_n x_0^n$ converge to 0; in particular, $|a_n x_0^n| \leq C$. Therefore

$$|a_n x^n| \leq C|x/x_0|^n,$$

and $\sum a_n x^n$ converges absolutely by comparison with the geometric series $\sum C|x/x_0|^n$.

Next we prove uniform convergence on $[-R_0, R_0]$. For some leeway, choose R_1 between R_0 and R. Since the series converges for $x = R_1$, the terms $a_n R_1^n$ converge to 0; in particular, $|a_n R_1^n| \leq C$. Hence for all $|x| \leq R_0$,

$$|a_n x^n| \leq C(R_0/R_1)^n.$$

To apply the Weierstrass M-test, let $M_n = C(R_0/R_1)^n$; then $\sum M_n$ is a convergent geometric series. By Weierstrass, $\sum a_n x^n$ converges uniformly on $[-R_0, R_0]$.

In particular, for $[a, b] \subset (-R, R)$, by Theorem 16.5, the integral of the series equals the limit of the integrals of the partial sums, which of course equals the limit of the partial sums of the integrals of the terms, which equals the series of the integrals of the terms.

For example,

$$\int_0^{1/2} (1 + x + x^2 + x^3 + \cdots)\,dx = \left[x + \frac{x^2}{2} + \frac{x^3}{3} + \frac{x^4}{4} + \cdots\right]_0^{1/2}$$
$$= \frac{1}{2} + \frac{1}{2^2 2} + \frac{1}{2^3 3} + \frac{1}{2^4 4} + \cdots .$$

□

20.5. Hadamard formula. *The radius of convergence R of a power series $\sum a_n x^n$ is given by the formula*

$$R = \frac{1}{\limsup |a_n|^{1/n}},$$

under the agreement that $1/0 = \infty$ and $1/\infty = 0$.

Proof. Using the Root test (19.5), we have $\rho = \limsup |a_n x^n|^{1/n} = |x|\frac{1}{R}$. Hence by the Root test, the series converges absolutely if $|x|\frac{1}{R} < 1$, i.e., if $|x| < R$, and diverges if $|x|\frac{1}{R} > 1$, i.e., if $|x| > R$. Therefore, R is the radius of convergence. □

20.6. Corollary. *Given a power series $\sum a_n x^n$, the "derived series" $\sum n a_n x^{n-1}$ has the same radius of convergence.*

Proof. Multiplication by x (which does not depend on n) of course yields a series with the same radius of convergence: $\sum n a_n x^n$. Since $\lim n^{1/n} = 1$, by Hadamard's theorem the radius of convergence is the same as for $\sum a_n x^n$. □

20.7. Differentiation of power series. *Inside the radius of convergence, a power series may be differentiated term by term:*

$$\left(\sum a_n x^n\right)' = \sum n a_n x^{n-1}.$$

Proof. On an interval $[-R_0, R_0] \subset (-R, R)$, $\sum a_n x^n$ converges uniformly to some $f(x)$, and the derived series $\sum n a_n x^{n-1}$ converges uniformly to some $g(x)$. Since g can be integrated term by term, $\int_{-R_0}^{x} g = f(x) - f(-R_0)$. By the Fundamental Theorem of Calculus, $g(x) = f'(x)$, as desired. □

20.8. Corollary. *A real-analytic function is infinitely differentiable (C^∞).*

20.9. Corollary. *Inside the radius of convergence, you can antidifferentiate a power series term by term.*

20.10. Taylor's formula. *If $f(x) = \sum a_n x^n$, then $a_n = \frac{f^{(n)}}{n!}$, where $f^{(n)}$ denotes the nth derivative.*

Proof.

$$\begin{aligned}
f(x) &= a_0 + a_1x + a_2x^2 + a_3x^3 + \cdots, & f(0) &= a_0;\\
f'(x) &= \quad a_1 + 2a_2x + 3a_3x^2 + \cdots, & f'(0) &= 1a_1;\\
f''(x) &= \quad 2a_2 + 3\cdot 2a_3x + \cdots, & f''(0) &= 2a_2;\\
f'''(x) &= \quad 3!\,a_3 + \cdots, & f'''(0) &= 3!\,a_3;
\end{aligned}$$

and so on. □

Caveat. Given a C^∞ function f, Taylor's formula yields an associated Taylor Series. Even if the domain of f is all of $\mathbb{R}$, the Taylor Series could have radius of convergence 0, and even if the series converges, it need not converge to the original function f. In a more advanced course, you will see examples of this strange behavior.

20.11. Power series about $c \neq 0$. Power series $\sum a_n x^n$ are best when x is near 0 ($x \approx 0$). For $x \approx c$, we could use $\sum a_n(x-c)^n$. There are only minor changes to the theory. The interval of convergence now goes from $c - R$ to $c + R$, and Taylor's formula becomes

$$a_n = \frac{f^{(n)}(c)}{n!}. \tag{1}$$

For example, $f(x) = 1/x$ has no power series about 0 because it is not even defined there. However, in terms of $u = x - 1$,

$$\begin{aligned}
\frac{1}{x} = \frac{1}{1+u} &= 1 - u + u^2 - u^3 + u^4 - \ldots \quad (\text{for } |u| < 1)\\
&= 1 - (x-1) + (x-1)^2 - (x-1)^3 + (x-1)^4 - \ldots \quad (\text{for } |x-1| < 1).
\end{aligned} \tag{2}$$

Exercises 20

1. a. Prove that

$$x - \frac{x^3}{3!} + \frac{x^5}{5!} - \frac{x^7}{7!} + \frac{x^9}{9!} - \cdots$$

has radius of convergence $R = \infty$. (Hint: Use the Ratio test 19.4.)

b. TRUE/FALSE. It follows that the series converges absolutely and uniformly on every bounded interval $[-R_0, R_0]$.

2. Consider $f(x) = \sum_{n=1}^{\infty} \frac{x^n}{n^2}$.

a. Use the Ratio test to show that the radius of convergence $R = 1$.

b. Use the Hadamard formula to show that $R = 1$.

c. It follows that the series converges uniformly on every interval $[-R_0, R_0] \subset (-1, 1)$. Use the Weierstrass M-test to show that it actually converges uniformly on $[-1, 1]$.

3. Continuing Exercise 2, consider $g(x) = xf'(x)$.

a. Compute that $g(x) = \sum_{n=1}^{\infty} \frac{x^n}{n}$.

b. TRUE/FALSE. By Corollary 20.6, $g(x)$ has the same radius of convergence $R = 1$.

c. Use the Ratio test and Hadamard formula to show that $R = 1$.

d. It follows that the series for $g(x)$ converges absolutely for $|x| < 1$. At the endpoints of the interval of convergence 1 and -1, does the series converge absolutely, converge conditionally, or diverge?

4. Find the Taylor Series and its radius of convergence for

a. $f(x) = e^x$;

b. $f(x) = x^2$.

5. *Leibniz's formula for* π.

a. Justify that for $|x| < 1$,

$$\frac{1}{1+x^2} = 1 - x^2 + x^4 - x^6 + x^8 - \cdots .$$

b. Justify that for $|x| < 1$,

$$\tan^{-1} x = x - \frac{x^3}{3} + \frac{x^5}{5} - \frac{x^7}{7} + \frac{x^9}{9} - \cdots .$$

(Note that it does not suffice to show that this is the Taylor series for $\tan^{-1} x$; see the Caveat after 20.10.)

c. Assuming that part *b* also holds for $x = 1$ (as it does), deduce Leibniz's famous formula for π:

$$\pi = \frac{4}{1} - \frac{4}{3} + \frac{4}{5} - \frac{4}{7} + \frac{4}{9} - \cdots .$$

6. Check Taylor's formula 20.11(1) for the power series 20.11(2).

7. Find a series for $1/x$ in powers of $u = x - 2$ by noting that

$$\frac{1}{x} = \frac{1}{2+u} = \frac{1}{2}\,\frac{1}{1+u/2}$$

and using the formula for a geometric series. Check Taylor's formula 20.11(1).

Chapter 21

The Exponential Function

It is no small feat to define a new function.

There are several ways to define the exponential function $\exp(x)$ or e^x and derive its properties.

The most obvious definition is just as powers of the constant e. For that approach, first you need to define e, perhaps as $\lim(1 + 1/n)^n$. Then $e^2 = e \times e$, $e^n = e \times e \times \cdots \times e$ (n times), $e^{-n} = 1/e^n$, a rational power $e^{p/q}$ is the qth root of e^p, but real powers are more complicated, and the various properties of e^x do not follow easily.

Rigorous calculus books define the natural logarithm first as an integral, and then define the exponential function as the inverse.

Here we follow more sophisticated texts by defining the exponential function by a power series. You may be surprised at how easily all the desired properties follow.

21.1. Definition. Define the exponential function

$$\exp(x) = \sum_{n=0}^{\infty} \frac{x^n}{n!} = 1 + x + \frac{x^2}{2!} + \frac{x^3}{3!} + \cdots .$$

Define $e = \exp(1) = \sum \frac{1}{n!} \approx 2.718281828$.

21.2. Proposition. *The series defining* $\exp(x)$ *converges absolutely on all of* $\mathbb{R}$*, uniformly on every bounded interval. The function is* C^∞ *and may be integrated and differentiated term by term. It is its own derivative.*

$\exp(0) = 1$ *and* $\exp(x)$ *is positive for* $x \geq 0$.

Proof. For the Ratio test (19.4), $\rho = \lim |x|/n = 0$; so the radius of convergence is ∞ and by Theorem 20.4 the series converges uniformly on bounded intervals. By 20.7–20.9, $\exp(x)$ is C^∞ and may be integrated and differentiated term by term, to yield $\exp'(x) = 0 + 1 + x + x^2/2! + \cdots = \exp(x)$.

It follows immediately from the definition that $\exp(0) = 1$ and $\exp(x)$ is positive for $x \geq 0$. □

21.3. Proposition. $\exp(-b) = \frac{1}{\exp(b)}$. *In particular,* $\exp(-1) = 1/e$, *and* $\exp(x)$ *is positive for all* x.

Proof. This is the part of the development where you might think you would have trouble. Dividing 1 by an infinite series is quite a long division problem. But it turns out that there is a slick proof. Notice that by the product rule

$$\frac{d}{dx}(\exp(x)\exp(-x)) = \exp(x)\exp(-x) - \exp(x)\exp(-x) = 0.$$

Hence $\exp(x)\exp(-x)$ is a constant function: $\exp(x)\exp(-x) = C$. Plugging in $x = 0$, we see that $C = 1$. Hence $\exp(-x) = 1/\exp(x)$, as desired. □

21.4. Proposition. $\exp(a + b) = \exp(a)\exp(b)$.

Proof. Since by the quotient rule

$$\frac{d}{dx}\frac{\exp(a+x)}{\exp(x)} = \frac{\exp(a+x)\exp(x) - \exp(a+x)\exp(x)}{\exp^2(x)} = 0,$$

$\exp(a+x) = C\exp(x)$. Plugging in $x = 0$ yields $C = \exp(a)$, as desired. □

21.5. Corollary. $\exp(n) = e^n$; $\exp(p/q) = \sqrt[q]{e^p}$.

Proof. By Proposition 21.4, $\exp(n) = \exp(1)\exp(1)\ldots\exp(1) = e^n$. Likewise, $(\exp(p/q))^q = \exp(p) = e^p$, which implies that $\exp(p/q) = e^{p/q}$. If we now define e^r as $\exp(r)$ for all real numbers, this definition will agree with any others by continuity. □

21.6. Corollary.

(1) $\exp(x)$ *is strictly increasing.*

(2) *For every* n, *there is an* $\varepsilon_n > 0$, *such that for* $x \geq 0$, $\exp(x) > \varepsilon_n x^n$.

(3) *Given* $C > 0$ *and* n, *for* x *large,* $\exp(x) > Cx^n$.

Proof. Conclusion (1) follows because the derivative $\exp(x)$ is positive. Conclusion (2) follows from the defining series, with $\varepsilon_n = 1/n!$. In particular, given $C > 0$, for $x \geq 0$, $\exp(x) > (1/(n+1)!)x^{n+1}$, which is greater than Cx^n once $x > C(n+1)!$. □

21.7. The natural logarithm. It follows from Corollary 21.6(1) that $\exp(x)$ is a bijective map from $\mathbb{R}$ onto the positive reals. The inverse function from the positive reals onto $\mathbb{R}$ is called the natural logarithm function, written $\ln x$ in calculus books and $\log x$ in more advanced mathematics. Its properties, such as $(\log x)' = 1/x$, follow from the properties of $\exp(x)$. See Exercise 4.

21.8. Complex exponentials. Although we have dealt only with real series, all of our results hold for series of complex numbers $z = x + iy$, including the definition and properties of $\exp(x)$. Here $i^2 = -1$, $i^3 = -i$, $i^4 = 1, \ldots$.

Exercises 21

1. *Development of the sine and cosine functions.* Define

$$\cos x = 1 - \frac{x^2}{2!} + \frac{x^4}{4!} - \frac{x^6}{6!} + \frac{x^8}{8!} - \cdots,$$
$$\sin x = x - \frac{x^3}{3!} + \frac{x^5}{5!} - \frac{x^7}{7!} + \frac{x^9}{9!} - \cdots.$$

a. Check that the radii of convergence are ∞, so that these functions are defined for all x.

b. Show that $\cos(-x) = \cos x$ and that $\sin(-x) = -\sin(x)$.

c. Show that $(\sin x)' = \cos x$ and that $(\cos x)' = -\sin x$.

d. Prove that $\sin^2 x + \cos^2 x = 1$.

Hint: First show the derivative 0, so that $\sin^2 x + \cos^2 x = C$. Then use $x = 0$ to find C.

Note that this implies that $|\sin x| \leq 1$ and $|\cos x| \leq 1$.

2. By plugging into the series for e^x, verify Euler's identity

$$e^{i\theta} = \cos\theta + i\sin\theta.$$

3. Use Euler's identity and the fact that $e^{i(a+b)} = e^{ia}e^{ib}$ to deduce that

$$\sin(a+b) = \sin a\cos b + \cos a\sin b,$$
$$\cos(a+b) = \cos a\cos b - \sin a\sin b.$$

4. Since the exponential function has a nonvanishing (never 0) continuous derivative, it follows from the Inverse Function Theorem (which did not quite make it into this book) that its inverse, $\log x$, is continuously differentiable.

a. If $y = \log x$, then $x = e^y$. Differentiate implicitly to deduce that

$$(\log x)' = 1/x.$$

b. Prove that $\log cx = \log c + \log x$.

5. Obtain a power series for $\log x$ in powers of $x - 1$ by integrating 20.11(2). Check Taylor's formula (20.11(1)) for this series.

6. Prove that $e = \lim_{n\to\infty}(1 + 1/n)^n$. (This is sometimes used as the definition of e.)

Hint: It suffices to show that

$$\log(1 + 1/n)^n = n \log(1 + 1/n) \to 1.$$

By the definition of the derivative,

$$\log'(1) = \lim_{\Delta x \to 0} \frac{\log(1 + \Delta x)}{\Delta x}.$$

Therefore, taking $\Delta x = 1/n$,

$$n \log(1 + 1/n) \to \log'(1) = 1.$$

Chapter 22

Volumes of n-Balls and the Gamma Function

There is a wonderful formula (22.5) for the volume of the ball $B(0,r)$ in $\mathbb{R}^n$:

$$\operatorname{vol} B(0,r) = \frac{\pi^{n/2}}{(n/2)!} r^n.$$

For example, the "volume" or area of a ball in $\mathbb{R}^2$ is $\frac{\pi^{2/2}}{(2/2)!}r^2 = \pi r^2$. The volume of a ball in $\mathbb{R}^4$ is $\frac{1}{2}\pi^2 r^4$. However, according to this formula, the "volume" or length of a ball in $\mathbb{R}^1$, which should be $2r$, is $\frac{\pi^{1/2}}{(1/2)!}r$. What is $(1/2)!$? For our formula to work, we would need

$$(1/2)! = \sqrt{\pi}/2.$$

Fortunately, there is a nice extension of $(x-1)!$ from integers to real numbers greater than 1, called the Gamma Function $\Gamma(x)$.

22.1. Definition. For $x \geq 1$, define the *Gamma Function* $\Gamma(x)$ by

$$\Gamma(x) = \int_0^\infty e^{-t} t^{x-1}\, dt.$$

This integral converges because e^t grows faster than every power of t.

22.2. Proposition. $\Gamma(x+1) = x\,\Gamma(x)$.

Proof. Using integration by parts, $\int u\,dv = uv - \int v\,du$, with $u = t^x$, $dv = e^{-t}dt$, $du = xt^{x-1}$, $v = -e^{-t}$,

$$\Gamma(x+1) = \int_0^\infty e^{-t}t^x\, dt = -e^{-t}t^x|_0^\infty + \int_0^\infty xe^{-t}t^{x-1}\, dt = 0 + x\,\Gamma(x). \quad \square$$

22.3. Proposition. *For every nonnegative integer n, $\Gamma(n+1) = n!$.*

We can use this relationship to extend the definition of the factorial function to every nonnegative real number by $x! = \Gamma(x+1)$.

Proof by induction. First we note that for $n = 0$,

$$\Gamma(0+1) = \Gamma(1) = \int_0^\infty e^{-t}\,dt = -e^{-t}|_0^\infty = -0 - (-1) = 1 = 0!.$$

Second, assuming the result for smaller values of n, we see by Proposition 22.2 that $\Gamma(n+1) = n\,\Gamma(n) = n(n-1)! = n!$. □

22.4. Proposition. $\Gamma(1/2) = \sqrt{\pi}$.

Proof. By making the substitution $t = s^2$, $dt = 2s\,ds$, we see that

$$\Gamma(1/2) = \int_0^\infty e^{-t}t^{-1/2}\,dt = \int_0^\infty e^{-s^2}s^{-1}\,2s\,ds = \int_{-\infty}^\infty e^{-s^2}\,ds.$$

Now comes the famous trick—multiplying $\Gamma(1/2)$ by itself and switching to polar coordinates with $dA = r\,dr\,d\theta$:

$$\begin{aligned}
\Gamma(1/2)\,\Gamma(1/2) &= \int_{-\infty}^\infty e^{-s^2}\,ds \int_{-\infty}^\infty e^{-u^2}\,du \\
&= \int_{-\infty}^\infty e^{-s^2} \int_{-\infty}^\infty e^{-u^2}\,du\,ds \\
&= \iint e^{-s^2}e^{-u^2}\,du\,ds = \iint e^{-(s^2+u^2)}\,du\,ds \\
&= \int_{\theta=0}^{2\pi}\int_{r=0}^\infty e^{-r^2}\,r\,dr\,d\theta = 2\pi\left[-\frac{1}{2}e^{-r^2}\right]_0^\infty = \pi.
\end{aligned}$$

Therefore $\Gamma(1/2) = \sqrt{\pi}$.

In particular, $(1/2)! = \Gamma(3/2) = (1/2)\,\Gamma(1/2) = \sqrt{\pi}/2$, as we hoped. Note too that our formula gives for the volume of a ball in $\mathbb{R}^3$:

$$\operatorname{vol} B(0,r) = \frac{\pi^{3/2}}{(3/2)!}r^3 = \frac{\pi^{3/2}}{(3/2)\sqrt{\pi}/2}r^3 = \frac{4}{3}\pi r^3,$$

the correct, familiar formula. □

22.5. Proposition. *The volume of a ball in $\mathbb{R}^n$ is given by*

$$\operatorname{vol} B(0,R) = \frac{\pi^{n/2}}{(n/2)!}R^n.$$

Proof. We haven't defined volume; we will assume here that it can be computed as in calculus, by integrating over slices. We've already checked this formula for $n = 1$ and $n = 2$. We'll prove it by induction, assuming the result for $n - 2$. (By jumping two dimensions at a time, we can use a much simpler polar coordinates proof.) We view the nD ball $\{x_1^2 + x_2^2 + x_3^2 + \cdots + x_n^2 \leq R^2\}$ as a 2D ball $\{r^2 = x_1^2 + x_2^2 \leq R^2\}$ of $(n-2)$D balls of radius s satisfying $s^2 = x_3^2 + \cdots + x_n^2 = R^2 - r^2$ and volume (by induction)

$$\frac{\pi^{(n-2)/2}}{((n-2)/2)!} s^{n-2} = \frac{\pi^{(n/2)-1}}{((n/2)-1)!} s^{n-2}.$$

Hence,

$$\begin{aligned}
\operatorname{vol} B(0,R) &= \iint \frac{\pi^{(n/2)-1}}{((n/2)-1)!} s^{n-2}\, dA \\
&= \frac{\pi^{(n/2)-1}}{((n/2)-1)!} \int_{\theta=0}^{2\pi} \int_{r=0}^{R} (R^2 - r^2)^{(n-2)/2} r\, dr\, d\theta \\
&= \frac{\pi^{(n/2)-1}}{((n/2)-1)!} 2\pi \left[-\frac{1}{2} \frac{(R^2 - r^2)^{n/2}}{n/2} \right]_0^R \\
&= \frac{\pi^{(n/2)-1}}{((n/2)-1)!} 2\pi \frac{1}{2} \frac{R^n}{n/2} = \frac{\pi^{n/2}}{(n/2)!} R^n.
\end{aligned}$$

□

22.6. Corollary. *The volume of the $(n-1)$D sphere in $\mathbb{R}^n$ is given by*

$$\operatorname{vol} S(0,R) = n \frac{\pi^{n/2}}{(n/2)!} R^{n-1}.$$

For example, the circumference of a circle in $\mathbb{R}^2$ is given by $2\frac{\pi}{1}R = 2\pi R$.

Proof. The volume of the sphere is just the derivative of the volume of the ball (the greater the surface area, the faster the volume of the ball increases). □

22.7. Stirling's Approximation. The Gamma function may be used to derive an excellent asymptotic approximation to $n!$

$$n! \sim \sqrt{2\pi n}(n/e)^n.$$

(Technically this asymptotic symbol $\sim$ means that as $n \to \infty$, $\frac{\sqrt{2\pi n}(n/e)^n}{n!} \to 1$.) It follows that

$$e^n \ll n! \ll n^n.$$

(For the definition of rates of growth, see 3.7.)

Exercises 22

1. a. Use Propositions 22.2 and 22.4 to find a numerical value for $(3/2)! = \Gamma(5/2)$.

b. Check the formula for the volume of the ball $B(0, r)$ in $\mathbb{R}^3$.

2. What is the volume of the ball $B(0, r)$ in $\mathbb{R}^5$? in $\mathbb{R}^6$?

3. What is the area or volume of the sphere (surface of the ball) $S(0, r)$ in $\mathbb{R}^3$? in $\mathbb{R}^4$? in $\mathbb{R}^5$?

4. Use Stirling's Approximation and your calculator to estimate that $\log_{10} 1000! \approx 2567$, so that roughly $1000! \approx 10^{2567}$. Try to estimate $1000!$ by any other means.

5. Use Stirling's Approximation to prove that $e^n \ll n! \ll n^n$.

Part IV

Fourier Series

Chapter 23

Fourier Series

23.1. Sines and cosines. The functions $\sin nx$ and $\cos nx$ on $[-\pi, \pi]$ are beautiful oscillations representing pure tones. A majestic orchestral chord, represented by a much more complicated function, is made up of many such pure tones. The remarkable underlying mathematical fact is that *every* smooth function on $[-\pi, \pi]$ can be decomposed as an infinite series in terms of sines and cosines, called its Fourier series. This is in strong contrast to the fact that only real-analytic functions are given by Taylor series in powers of x. Apparently, for decomposition, sines and cosines work much better than powers of x. There is good reason for studying sines and cosines so much, starting in high school.

23.2. Theorem (Fourier series). *Every continuous, piecewise differentiable function $f(x)$ on $[-\pi, \pi]$ is given by a Fourier series*

$$f(x) = A_0 + \sum_{n=1}^{\infty}(A_n \cos nx + B_n \sin nx), \tag{1}$$

where

$$\begin{aligned} A_0 &= \frac{1}{2\pi}\int_{-\pi}^{\pi} f(x)\,dx, \\ A_n &= \frac{1}{\pi}\int_{-\pi}^{\pi} f(x)\cos nx\,dx, \\ B_n &= \frac{1}{\pi}\int_{-\pi}^{\pi} f(x)\sin nx\,dx, \end{aligned} \tag{2}$$

except that at the endpoints the series converges to the average of $f(-\pi)$ and $f(\pi)$.

Outside $[-\pi, \pi]$, the Fourier series just keeps repeating every 2π, so it doesn't always equal f, *unless* f is also 2π periodic.

The proof of Theorem 23.2 is just beyond the scope of this text, although Theorem 25.1 at least proves that the series *converges* if f is smooth; it takes more work to prove that it *converges to* f.

23.3. The coefficients. The above formulas 23.2(2) for the *Fourier coefficients* A_n and B_n, due to Euler, deserve comment. First note that A_0 is just the average value of $f(x)$, about which f varies or oscillates. The other formulas are best understood by comparison with formulas for the coefficients of a vector

$$\mathbf{v} = v_1 e_1 + v_2 e_2 + \cdots + v_N e_N \tag{1}$$

in $\mathbb{R}^N$, expressed as a linear combination of standard basis vectors e_n. Such basis vectors are orthonormal, that is, $e_m \cdot e_n = 0$, except that $e_m \cdot e_m = 1$. It follows immediately that the coefficients v_n satisfy

$$v_n = \mathbf{v} \cdot e_n, \tag{2}$$

by just taking the dot product of each side of (1) with e_n.

Analogously, on the space of continuous functions on $[-\pi, \pi]$, one can define a dot product by

$$f \cdot g = \frac{1}{\pi} \int_{-\pi}^{\pi} f(x)\, g(x)\, dx, \tag{3}$$

integrating all products of values instead of just summing products of components. We want to write everything in terms of the functions $\cos nx$ and $\sin nx$, which turn out to be orthonormal functions for this dot product (see Exercise 1). The formulas for the Fourier coefficients 23.2(2) just say that each Fourier coefficient is obtained by dotting the function $f(x)$ with the associated orthonormal function, just like the corresponding formula (2) for vectors.

23.4. Example. For example, consider the piecewise smooth function

$$f(x) = \begin{cases} 0 & \text{if } -\pi \le x \le 0, \\ x(\pi - x) & \text{if } 0 \le x \le \pi \end{cases}$$

of Figure 23.1. Exercise 3 uses 23.2(2) to compute the Fourier series:

$$f(x) = \frac{\pi^2}{12} - \sum_{n \text{ even}} \frac{2}{n^2} \cos nx + \sum_{n \text{ odd}} \frac{4}{\pi n^3} \sin nx. \tag{1}$$

(See Figure 23.2.)

Fourier series have many serious real-world applications, as in the upcoming Chapter 24. Meanwhile, here is a more amusing application. The

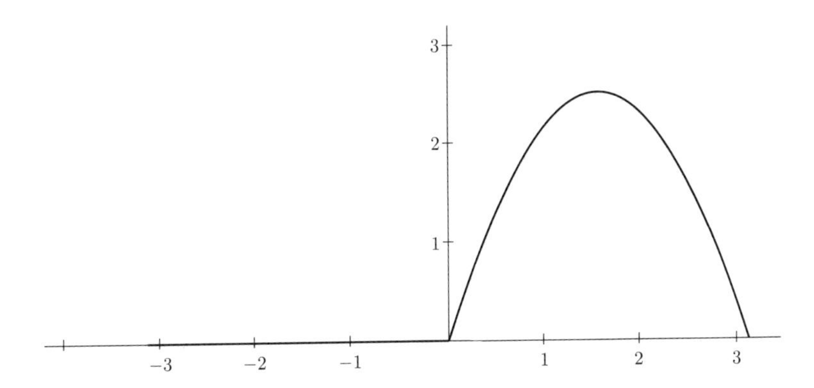

Figure 23.1. The piecewise smooth function $f(x) = x(\pi - x)$ for $0 \leq x \leq \pi$.

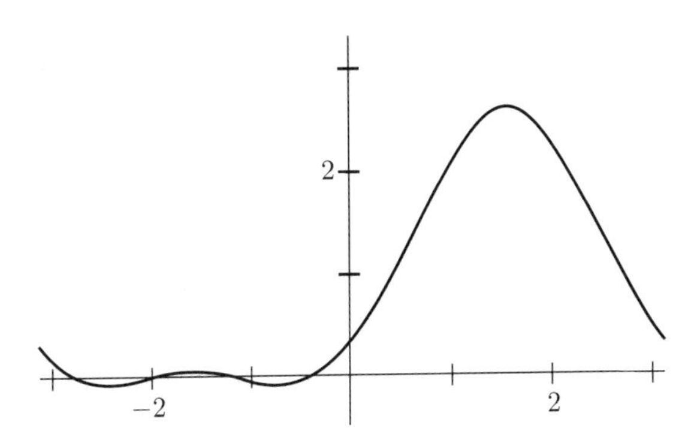

Figure 23.2. Sum of the first few terms of the Fourier series for the function of Figure 23.1.

following Proposition 23.5 has many proofs; in ours, Fourier series make an unexpected appearance. The proposition gives the exact sum of the previously mysterious series $\sum(1/n^2)$. By the p-test 18.4, we knew that this series converged, and now we finally find out the surprising answer to what it converges to (how did that π get in there?).

23.5. Proposition. *The series* $\sum(1/n^2)$ *converges to* $\pi^2/6$.

Proof. Plugging $x = 0$ into the Fourier series 23.4(1) yields:

$$0 = \frac{\pi^2}{12} - \left(\frac{2}{2^2} + \frac{2}{4^2} + \frac{2}{6^2} + \dots\right).$$

Multiplying by 2 yields

$$0 = \frac{\pi^2}{6} - \left(\frac{1}{1^2} + \frac{1}{2^2} + \frac{1}{3^2} + \dots\right)$$

so that $\sum(1/n^2) = \pi^2/6$. □

23.6. Perspective. Since a Fourier series is determined by the coefficients, Fourier series provide a way to encode a function by two sequences A_n, B_n of numbers with physical significance corresponding to "frequencies." For a violin string, the predominant frequencies are nice multiples of each other, called harmonics, and produce a rich, harmonious sound. In many applications, it is often the case that certain frequencies dominate, and the function can be well approximated by finitely many terms. Such mathematics can provide a way to encode or compress complicated data, sounds, or images. JPEG is exactly such a method for efficient storage of images.

Dolby stereo filters out hissing noise by avoiding the frequencies responsible for such noise.

The next chapter (24) will show how Fourier series can be used to solve important differential equations.

23.7. The sine series. If $f(x)$ is an odd function, which means that $f(-x) = -f(x)$, it follows immediately (Exercise 5) from the formulas 23.2(2) for the coefficients that every A_n vanishes (equals 0), so that the Fourier series consists entirely of sines.

Every function $f(x)$ on $(0, \pi)$ can be extended to an odd function on $(-\pi, \pi)$, whose sine series holds for the original function $f(x)$ on $(0, \pi)$:

$$f(x) = \sum_{n=1}^{\infty} B_n \sin nx, \tag{1}$$

where

$$B_n = \frac{2}{\pi} \int_0^{\pi} f(x) \sin nx \, dx. \tag{2}$$

The new factor of 2 comes because we are integrating only from 0 to π; we wouldn't need it if we integrated the extension from $-\pi$ to π.

23.8. General intervals. Fourier series (23.2) immediately generalize from $[-\pi, \pi]$ to $[-L, L]$, just by replacing x with $\pi x / L$:

$$f(x) = A_0 + \sum_{n=1}^{\infty} \left(A_n \cos \frac{n\pi x}{L} + B_n \sin \frac{n\pi x}{L} \right), \tag{1}$$

where

$$
\begin{aligned}
A_0 &= \frac{1}{2L}\int_{-L}^{L} f(x)\,dx, \\
A_n &= \frac{1}{L}\int_{-L}^{L} f(x)\cos\frac{n\pi x}{L}\,dx, \qquad (2) \\
B_n &= \frac{1}{L}\int_{-L}^{L} f(x)\sin\frac{n\pi x}{L}\,dx,
\end{aligned}
$$

Exercises 23

1. Verify that the functions $\cos x$ and $\sin x$ are orthonormal on $[-\pi, \pi]$ for the dot product 23.3(3).

2. What is the Fourier series for $f(x) = \sin x$ on $[-\pi, \pi]$?

3. Compute the Fourier series for Example 23.4 using 23.2(2).
Hint: Use integration by parts and notice that $\cos n\pi + 1$ is 0 if n is odd and 2 if n is even.

4. Show that the Fourier series for $f(x) = x$ $(-\pi \leq x \leq \pi)$ is given by

$$x = 2\left(\frac{1}{1}\sin x - \frac{1}{2}\sin 2x + \frac{1}{3}\sin 3x - \cdots\right).$$

Taking $x = \pi/2$, obtain Leibniz's amazing formula for π:

$$\pi = \frac{4}{1} - \frac{4}{3} + \frac{4}{5} - \frac{4}{7} + \cdots.$$

5. Show that if $f(x)$ is an odd function, which means that $f(-x) = -f(x)$, then the coefficients A_n vanish, and you get a Fourier series of sines. Similarly if $f(x)$ is an even function, which means that $f(-x) = f(x)$, then the coefficients B_n vanish, and you get a Fourier series of cosines.

6. Show that the Fourier series for the function

$$f(x) = \begin{cases} 0 & \text{for } -\pi \leq x \leq 0, \\ x & \text{for } 0 \leq x \leq \pi/2, \\ \pi - x & \text{for } \pi/2 \leq x \leq \pi, \end{cases}$$

is given by

$$f(x) = \frac{\pi}{8} - \frac{4}{\pi}\left(\frac{1}{2^2}\cos 2x + \frac{1}{6^2}\cos 6x + \frac{1}{10^2}\cos 10x + \cdots\right)$$
$$+ \frac{2}{\pi}\left(\frac{1}{1^2}\sin x - \frac{1}{3^2}\sin 3x + \frac{1}{5x^2}\sin 5x - \ldots\right).$$

Chapter 24

Strings and Springs

Chapter 24 provides classic applications of Fourier series to vibrations of a string and to oscillations of a spring.

24.1. The vibrating string. In 1755 Daniel Bernoulli found a general solution $y(x, t)$ to the differential equation for the small vertical displacement y of a vibrating string as a function of $0 \leq x \leq \pi$ and time t. We'll assume that the string has unit density and tension and is fixed at its endpoints. The simplest solution was

$$y(x, t) = \sin x \cos t,$$

the curve $y = \sin x$ oscillating in time, as in Figure 24.1.

Similarly there were simple higher frequency solutions

$$y(x, t) = \sin nx \cos nt,$$

oscillating at multiples of the fundamental frequency, as in Figure 24.2.

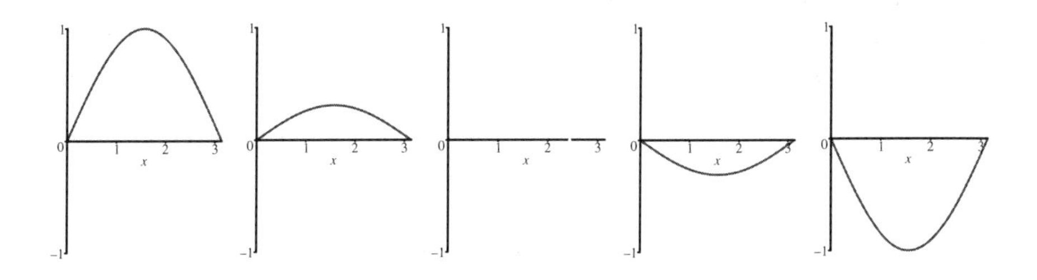

Figure 24.1. The simplest vibration of a string.

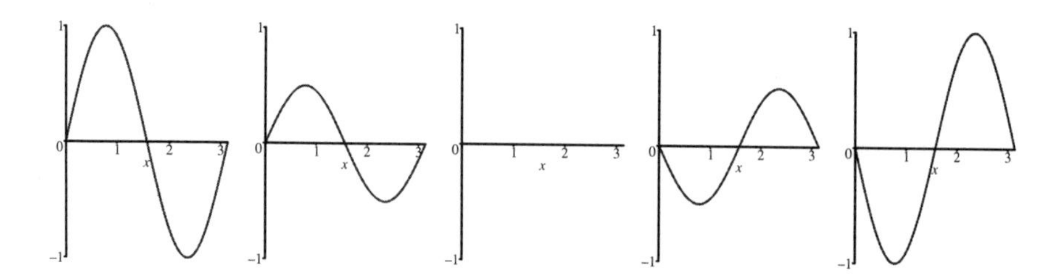

Figure 24.2. The second simplest vibration of a string.

Finite linear combinations of solutions are solutions. For the general solution, Bernoulli proposed infinite linear combinations:

$$y(x,t) = \sum_{n=1}^{\infty} B_n \sin nx \cos nt. \tag{1}$$

At time $t = 0$, the initial shape is given by

$$y(x,0) = \sum_{n=1}^{\infty} B_n \sin nx. \tag{2}$$

Many mathematicians asked, "What about other initial shapes?" The surprising answer was that *every* nice function is given as such a Fourier series 23.7(1). This Fourier series for the initial shape (2) determines the contribution of the various frequencies of oscillation in the solution (1).

To think about such questions in any detail as we began in Chapter 23 requires a basic understanding of real analysis. Bernoulli had no idea, and it took mathematicians such as Weierstrass over a hundred years to develop what you already know. Questions about the generality and convergence of Fourier series were a major impetus in this development.

24.2. Forced spring oscillations. The displacement $x(t)$ of a unit mass on a spring (now using x as the dependent variable) with spring constant ω^2 and force $-\omega^2 x$ satisfies Newton's famous law $F = ma$. The acceleration a is the second derivative d^2x/dt^2, which Newton wrote as $\ddot{x}$, using a dot for differentiation with respect to time t. Thus $F = ma$ becomes

$$\ddot{x} + \omega^2 x = 0, \tag{1}$$

with well-known general solution

$$x = c_1 \cos \omega t + c_2 \sin \omega t. \tag{2}$$

If a driving force $f(t)$ is applied, the differential equation becomes the much harder

$$\ddot{x} + \omega^2 x = f(t). \tag{3}$$

We suppose that the driving force $f(t)$ is periodic with period 2π and seek a nice periodic solution. For example, suppose that a pulse as in Example 23.4 is applied every 2π seconds. Assume that x is given by a Fourier series

$$x = A_0 + \sum_{n=1}^{\infty}(A_n \cos nt + B_n \sin nt).$$

Hoping to be able to use our knowledge of real analysis to justify our actions later, we differentiate the series term by term,

$$\ddot{x} = -\sum_{n=1}^{\infty} n^2 (A_n \cos nt + B_n \sin nt), \tag{4}$$

and plug the result into equation (3) along with the Fourier series for f from Example 23.4. Equating coefficients of like terms yields the following equations for the coefficients:

$$\begin{aligned} &A_0 = \pi^2/12\omega^2, \\ n \text{ even:}\quad &(\omega^2 - n^2)A_n = -2/n^2, \quad B_n = 0, \\ n \text{ odd:}\quad &A_n = 0, \qquad (\omega^2 - n^2)B_n = 4/\pi n^3. \end{aligned} \tag{5}$$

Hence,

$$x(t) = \frac{\pi^2}{12\omega^2} + \sum_{n \text{ even}} -\frac{2}{n^2(\omega^2 - n^2)} \cos nt + \sum_{n \text{ odd}} \frac{4}{\pi n^3(\omega^2 - n^2)} \sin nt, \tag{6}$$

a Fourier series for our answer.

Resonance. Notice that the largest terms correspond to n closest to ω, the natural frequency of the spring. Forced vibrations close to the natural frequency cause huge oscillations, in a phenomenon called *resonance.* In one notorious case in 1940, the Tacoma Narrows suspension bridge, excited by winds at its natural frequency, oscillated ever more wildly until it tore itself apart. (Check it out on the web.) According to legend, soldiers break formation before crossing a bridge lest their cadence produce hazardous resonance.

Justification. The justification will use some of our strongest tools, the Weierstrass M-test (20.2) and Theorem 16.5 on integrating series term by term (moving the integral inside the summation sign). Start by defining $\ddot{x}$ as what you get if you differentiate (6) twice term by term:

$$\ddot{x} = \sum_{n \text{ even}} \frac{2}{\omega^2 - n^2} \cos nt - \sum_{n \text{ odd}} \frac{4}{n(\omega^2 - n^2)} \sin nt. \tag{7}$$

Note that for n large, the absolute value of the nth term is bounded by $M_n = 2/n^2$. Hence by the Weierstrass M-test (20.2), this Fourier series converges uniformly. Therefore by Theorem 16.5, it can be integrated term by term, yielding (6) up to a linear term $C_1 + C_2 t$, thus justifying the original term-by-term differentiation. Moreover, any solution x whose Fourier series when so twice differentiated is uniformly convergent must by the above process be our solution (6). The equating of like terms will be justified by Exercise 25.2.

Exercises 24

1. Find the Fourier series 23.7(1) in terms of sines for the function

$$f(x) = \begin{cases} x & \text{for } 0 \leq x \leq \pi/2, \\ \pi - x & \text{for } \pi/2 \leq x \leq \pi. \end{cases}$$

2. A plucked string has initial shape $\varepsilon f(x)$ from Exercise 1. Give its solution 24.1(1) as a function of x and t.

3. Check that 24.2(2) satisfies 24.2(1).

4. Derive equations 24.2(5).

5. Show that a periodic solution to 24.2(3), for $f(t)$ a unit pulse from $t = 0$ to $t = \pi/2$ repeated with period 2π, is given by

$$x(t) = \frac{1}{4\omega^2} + \sum_{n=1}^{\infty} \frac{\sin \frac{n\pi}{2} \cos nt + \left(1 - \cos \frac{n\pi}{2}\right) \sin nt}{n\pi(\omega^2 - n^2)}.$$

6. Check that $y(x,t) = \sin nx \cos nt$ is a solution of the wave equation

$$\frac{\partial^2 y}{\partial t^2} = \frac{\partial^2 y}{\partial x^2}.$$

Chapter 25

Convergence of Fourier Series

Theorem 25.1 provides the promised sample of Fourier theory, relatively easy for you, utterly beyond Bernoulli and 18th Century mathematics.

25.1. Theorem. *Let f be a C^2 function on $[-\pi, \pi]$ with $f(\pi) = f(-\pi)$ and $f'(\pi) = f'(-\pi)$. Then the Fourier series for f converges uniformly.*

Proof. Let $C = \max\{f''(x)\}$. By integration by parts,

$$\begin{aligned} A_n &= \frac{1}{\pi}\int_{-\pi}^{\pi} f(x)\cos nx\,dx \\ &= \frac{1}{\pi}\frac{1}{n} f(x)\sin nx|_{-\pi}^{\pi} - \frac{1}{\pi}\frac{1}{n}\int_{-\pi}^{\pi} f'(x)\sin nx\,dx \\ &= 0 + \frac{1}{\pi}\frac{1}{n^2} f'(x)\cos nx|_{-\pi}^{\pi} - \frac{1}{\pi}\frac{1}{n^2}\int_{-\pi}^{\pi} f''(x)\cos nx\,dx \\ &= 0 + 0 - \frac{1}{\pi}\frac{1}{n^2}\int_{-\pi}^{\pi} f''(x)\cos nx\,dx. \end{aligned}$$

Hence,

$$|A_n \cos nx| \leq \frac{1}{\pi}\frac{1}{n^2}(2\pi)C = \frac{2C}{n^2}.$$

Similarly, $|B_n \sin nx| \leq 2C/n^2$. Therefore by the Weierstrass M-test (20.2), the Fourier series converges uniformly. □

25.2. Remarks. It takes a bit more work than we are prepared to do to show that the limit is actually f.

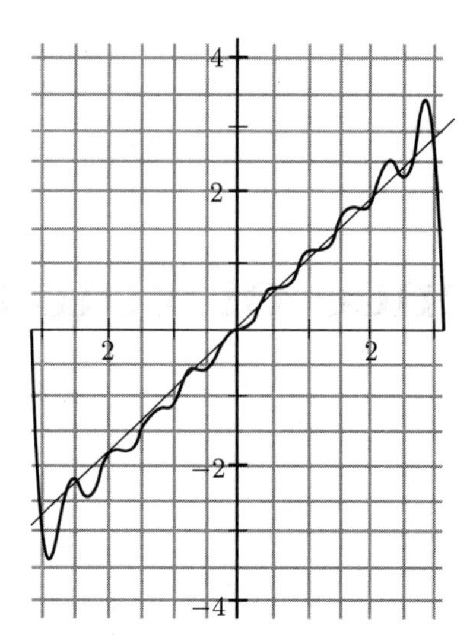

Figure 25.1. The Fourier approximations to $f(x) = x$ overshoot by about 9% at the endpoints, the so-called Gibbs phenomenon.

If $f(\pi) \neq f(-\pi)$, the Fourier series, which according to Theorem 23.2 converges to $(f(\pi) + f(-\pi))/2$ at the endpoints, thus has a limit discontinuous at the endpoints. Hence, it cannot converge uniformly because a uniform limit of continuous functions is continuous (Theorem 16.3). Worse, it actually *overshoots* as it approaches the endpoints by about a mysterious 9% of the difference $f(\pi) - f(-\pi)$, in what is called the *Gibbs phenomenon.* Figure 25.1 shows the sum of the first ten sine terms of the Fourier series for $f(x) = x$, which before coming back to 0 at the right endpoint overshoots the value of π by about 9% of 2π or about .57.

Exercises 25

1. In the proof of 25.1, verify that $|B_n \sin nx| \leq 2C/n^2$.

2. Prove that if a series of the form 23.2(1) converges uniformly to $f(x)$, then it must be the Fourier series for $f(x)$.

Hint: $\cos nx$ times the series converges uniformly to $f(x) \cos nx$, and hence may be integrated term by term.

3. Use Exercise 2 to solve Exercise 23.2 effortlessly.

Part V

The Calculus of Variations

Part V: The Calculus of Variations. Many problems in mathematics and in the physical and social sciences can be solved as max-min problems in huge, infinite-dimensional spaces of functions instead of in $\mathbb{R}^1$ or $\mathbb{R}^n$. Such infinite-dimensional calculus is called the calculus of variations. We'll consider various applications to geometry, physics, economics, and general relativity. The generalization of "$y' = 0$" is called Euler's equation.

Chapter 26

Euler's Equation

In first semester calculus, you seek the value of x to minimize some function $f(x)$. In multivariable calculus, you seek the values of $(x_1, x_2, \ldots, x_n)$ to minimize some function $f(x_1, x_2, \ldots, x_n)$ of several variables. In the calculus of variations, you seek a *function* $y = f(x)$ to minimize some cost, usually expressed as an integral involving x, y, and y':

$$\int_a^b F(x, y, y')dx, \tag{1}$$

where F is a function of three variables, which we will assume has continuous first and second partial derivatives.

Sample problem 26.6 below will minimize

$$\int_1^2 \frac{y'^2}{x} dx.$$

You might want to skim that example before seeing the theory.

The analog of setting the derivative equal to zero in first semester calculus is Euler's equation 26.5, a differential equation for $y = f(x)$, which you can solve to find the function $y = f(x)$ that you seek. Before deriving Euler's equation, we need a few lemmas.

26.1. Leibniz's Rule (special case). *Suppose that* $f(x,t)$ *and* $\frac{\partial f}{\partial t}$ *are continuous for* x *in* $[a,b]$ *and* t *in* $[t_0-\delta, t_0+\delta]$ *for some* $\delta > 0$. *Then*

$$\frac{d}{dt}\int_a^b f(x,t)dt\bigg|_{t=t_0} = \int_a^b \frac{\partial f}{\partial t}(x,t_0)dx.$$

Proof. As a continuous function on the compact set $[a,b]\times[t_0-\delta, t_0+\delta]$, $|\partial f/\partial t|$ has a maximum value c. Define $g(x) = c$. Now 26.1 follows immediately from Leibniz's Theorem 17.5. □

26.2. The Chain Rule. This is just a reminder of the following case of the chain rule from calculus. *Consider a function* $w = f(x,y,z)$, *with continuous partial derivatives* $\partial w/\partial x, \partial w/\partial y, \partial w/\partial z$. *Suppose that* x,y,z *are in turn differentiable functions of* t. *Then* w *is a differentiable function of* t *and*

$$\frac{dw}{dt} = \frac{\partial w}{\partial x}\frac{dx}{dt} + \frac{\partial w}{\partial y}\frac{dy}{dt} + \frac{\partial w}{\partial z}\frac{dz}{dt}.$$

26.3. Integration by Parts (special case). *Let* $f(x)$, $\eta(x)$ *(lower case Greek eta) be continuously differentiable functions on* $[a,b]$ *such that* $\eta(x)$ *vanishes (equals* 0*) at the endpoints. Then*

$$\int_a^b f\eta' = -\int_a^b f'\eta.$$

In other words, you can switch differentiation from one factor to another by changing the sign of the integral.

Proof. This follows immediately from the more familiar formula for integration by parts:

$$\int_a^b u\,dv = uv|_a^b - \int_a^b v\,du.$$

Indeed, let $u = f$, $v = \eta$, and note that the first term on the right vanishes because η vanishes at the endpoints. □

26.4. The Fundamental Lemma of the Calculus of Variations. *Let* f *be a continuous function on* $[a,b]$ *and suppose that*

$$\int_a^b f\eta = 0$$

for all smooth functions η *on* $[a,b]$ *vanishing about the endpoints. Then* f *is identically* 0.

Proof. Otherwise f is positive or negative somewhere, say positive, so that $f(c) > 0$ for some c in (a, b). By continuity, f is positive on a little interval $[c - \delta, c + \delta]$ inside (a, b). Let η be a smooth nonnegative "bump" function which is positive at c and 0 outside $[c - \delta, c + \delta]$, as in Figure 26.1. Then the integral of $f\ \eta$ has some positive contributions and no negative contributions, so that

$$\int_a^b f\eta > 0,$$

the desired contradiction. □

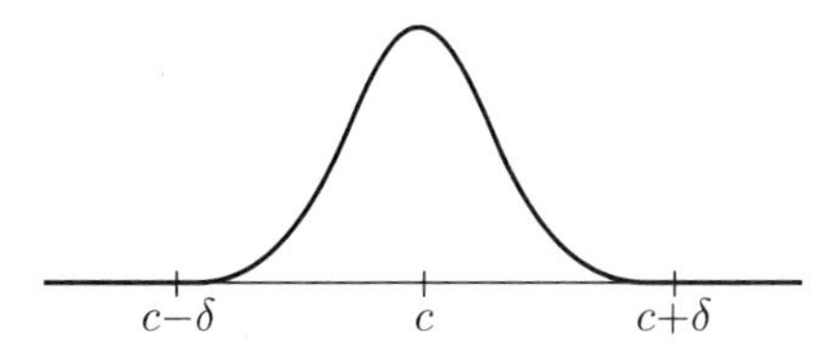

Figure 26.1. The "bump" function η is positive at c and 0 outside $[c - \delta, c + \delta]$.

26.5. Euler's Equation. *A local minimum (or maximum)* $y = f(x)$, *among all* C^2 *functions, of a cost*

$$\int_a^b F(x, y, y')dx,$$

given $f(a)$ *and* $f(b)$, *satisfies Euler's differential equation*

$$\frac{\partial F}{\partial y} = \frac{d}{dx}\frac{\partial F}{\partial y'}.$$

Proof. Obtain nearby functions $y(x) + \varepsilon\eta(x)$ by adding to $y(x)$ some small multiple of a smooth function $\eta(x)$ which vanishes at the endpoints a and b. The corresponding cost

$$C(\varepsilon) = \int_a^b F(x, y + \varepsilon\eta, y' + \varepsilon\eta')\, dx,$$

as a function of the single variable ε, has a local minimum (or maximum) at $\varepsilon = 0$. Hence from single-variable calculus,

$$0 = C'(0) = \frac{d}{d\varepsilon}\int_a^b F(x, y + \varepsilon\eta, y' + \varepsilon\eta')\, dx\Big|_{\varepsilon=0}.$$

Therefore at $\varepsilon = 0$,

$$\begin{aligned} 0 &= \int_a^b \frac{\partial}{\partial \varepsilon} F(x, y + \varepsilon\eta, y' + \varepsilon\eta')dx \quad \text{(by Leibniz's Rule 26.1)} \\ &= \int_a^b \frac{\partial F}{\partial y}\eta + \frac{\partial F}{\partial y'}\eta' \, dx \quad \text{(by the Chain Rule 26.2)} \\ &= \int_a^b \frac{\partial F}{\partial y}\eta - \frac{d}{dx}\frac{\partial F}{\partial y'}\eta \, dx \quad \text{(integration by parts)} \\ &= \int_a^b \left(\frac{\partial F}{\partial y} - \frac{d}{dx}\frac{\partial F}{\partial y'}\right)\eta \, dx. \end{aligned}$$

Since this holds for all smooth η vanishing at the endpoints (which we used for integration by parts), it follows from the Fundamental Lemma of the Calculus of Variations 26.4 that

$$\frac{\partial F}{\partial y} - \frac{d}{dx}\frac{\partial F}{\partial y'} = 0$$

as desired. □

26.6. Sample problem. Find a C^2 function $y = f(x)$ with $y(1) = 1$ and $y(2) = 2$ to minimize

$$\int_1^2 \frac{y'^2}{x} dx.$$

Two possibilities are shown in Figure 26.2. If the integrand did not have that x in the denominator, the minimizer would turn out to be $y = x$, with y' always 1, to avoid the heavy penalties that y'^2 imposes on higher values of y' (see Exercise 1). The x in the denominator makes it better to have smaller values of y' at first and larger values later. Maybe the solution should be quadratic or cubic or exponential? Before continuing, make your own guess.

Solution. Assuming that there is a minimizer y among C^2 functions with the given endpoint conditions, Euler's Equation 26.5 tells us that

$$\frac{\partial F}{\partial y} = \frac{d}{dx}\frac{\partial F}{\partial y'}.$$

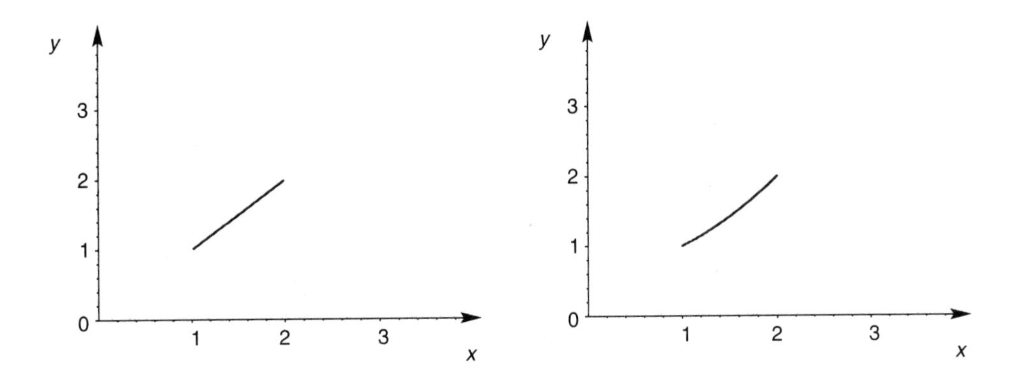

Figure 26.2. Two candidate solutions for Sample Problem 26.6.

Here, $F(x, y, z) = z^2/x$, a nice smooth function (at least for $1 \leq x \leq 2$), with $\partial F/\partial y = 0$ and $\partial F/\partial z = 2z/x$. Hence Euler's equation becomes

$$(1) \qquad 0 = \frac{d}{dx}\frac{2y'}{x} = \frac{2y''x - 2y'}{x^2}, \quad y''x = y'.$$

The method from differential equations for solving an equation like this which does not mention y explicitly is to substitute $y' = p$, so $y'' = dp/dx$, to make the equation:

$$\frac{dp}{dx}x = p,$$
$$\frac{dp}{p} = \frac{dx}{x},$$
$$\log|p| = \log|x| + C,$$
$$|p| = e^{\log|x|+C} = e^{\log|x|}e^C = C'x,$$
$$(2) \qquad y' = p = ax \quad (a = \pm C')$$
$$y = \frac{1}{2}ax^2 + b.$$

Substituting in the endpoint conditions $y(1) = 1$ and $y(2) = 2$ and solving for a and b yields

$$y = \frac{1}{3}x^2 + \frac{2}{3}.$$

Did you guess that the solution would be quadratic?

Incidentally, we could have taken a shortcut. Immediately from its derivative vanishing by (1), we could have concluded that $2y'/x$ is constant, i.e., $y' = ax$, equation (2). This is called the First Integral, as we'll see in the next chapter.

Remarks In our solution, we needed to assume that there was a C^2 minimizer. In general, the minimizer may not exist or it may not be C^2. Over the centuries, much work in real analysis has been devoted to finding the right hypotheses that guarantee that a solution exists and is C^2, so-called "existence and regularity." Fortunately, in many important applications, it seems obvious for physical reasons that a solution will exist and be nice.

Sometimes the Euler equation will have several solutions. There could be more than one minimizer, or more likely other relative minima, or perhaps saddle points, just like in multivariable calculus.

Exercises 26

1. Use Euler's Equation 26.5 to find a C^2 function $y = f(x)$ on $[1, 2]$ with $y(1) = 1$ and $y(2) = 2$ to minimize

$$\int_1^2 y'^2 dx.$$

(Assume that a C^2 minimizer exists.)

2. Use Euler's equation to find a C^2 function $y = f(x)$ on $[1, 2]$ with $y(1) = 3$ and $y(2) = 4$ that minimizes

$$\int_1^2 y'^2 - xy' - y\, dx.$$

(Assume that a C^2 minimizer exists.)

3. Find the general solution to Euler's equation for

$$\int_a^b y'^2/x^3\, dx.$$

4. Find the general solution to Euler's equation for

$$\int_a^b y'^2 - 8xy + x\, dx.$$

Chapter 27

First Integrals and the Brachistochrone Problem

Before consideration of the famous brachistochrone problem, which launched the whole subject of the calculus of variations, Proposition 27.1 gives a very useful technique for integrating Euler's equation in two special cases. Euler's Equation 26.5 involves two derivatives on the right and hence is a *second order* differential equation. Solving it requires two integrations, the first generally harder than the second. Proposition 27.1 provides a first integral in two cases: when the integrand F does not depend explicitly on x or on y.

27.1. First Integral. *If F does not depend explicitly on x, then Euler's Equation* 26.5 *implies that*

$$F - \frac{\partial F}{\partial y'} y' = C. \tag{1}$$

Conversely, if (1) holds, then either Euler's equation holds or $y' = 0$.

If F does not depend explicitly on y, then Euler's equation is equivalent to

$$\frac{\partial F}{\partial y'} = C. \tag{2}$$

Proof. Suppose that F does not depend explicitly on x, so that $\partial F/\partial x = 0$. Then (1) holds if and only if

$$\begin{aligned} 0 &= \frac{d}{dx}\left(F - \frac{\partial F}{\partial y'}y'\right) \\ &= \frac{\partial F}{\partial y}\frac{dy}{dx} + \frac{\partial F}{\partial y'}\frac{dy'}{dx} - \frac{d}{dx}\left(\frac{\partial F}{\partial y'}\right)y' - \frac{\partial F}{\partial y'}\frac{dy'}{dx} \\ &\qquad \text{(by the Chain rule on } F \text{ and the product rule on } \frac{\partial F}{\partial y'}y'\text{)} \\ &= \left(\frac{\partial F}{\partial y} - \frac{d}{dx}\frac{\partial F}{\partial y'}\right)y', \end{aligned}$$

which holds if and only if Euler's equation holds or $y' = 0$.

If F does not depend explicitly on y, then Euler's equation becomes

$$0 = \frac{d}{dx}\frac{\partial F}{\partial y'}$$

which is equivalent to (2). □

27.2. The Brachistochrone Problem. In 1697, John Bernoulli issued a challenge "to the most acute mathematicians of the entire world," but he had just one mathematician in mind: Newton, the great rival of his own teacher Leibniz for credit for the discovery of calculus. The brothers John and James Bernoulli were themselves friendly rivals, and John had originally issued the challenge to James. John had an elegant short solution based on the principle of optics. James came up with a longer but more general approach, which in time developed into the calculus of variations. After Leibniz himself had solved the problem (in about six months), he told John to send it on to Newton. Newton received it on January 29, 1697, after a day of work at the mint, and solved it before going to bed. He published his solution anonymously at the Royal Society, but John recognized the master at once and graciously admitted, "I recognize the lion by his paw":

Tanquam ex ungue leonem.

(Literally, "I recognize by his *claw* the lion.")

The problem was to find the shape of a slide down which a frictionless puck would slide in the least amount of time ("chrone") under the influence of gravity. A marvelous feature of the problem was that the solution was the already famous cycloid:

$$x = c(\varphi - \sin\varphi), \quad y = c(1 - \cos\varphi), \tag{1}$$

the path of a point on a rolling wheel, pictured in Figure 27.1 (we're measuring y downward to keep y nonnegative). It was famous as the shape of the path of a pendulum bob keeping perfect time. Standard pendula swing in circular arcs and speed up as the magnitude of the oscillations runs down. But if the pendulum bob is somehow constrained to follow a cycloid, then the time between beats remains constant, independent of how high the bob starts.

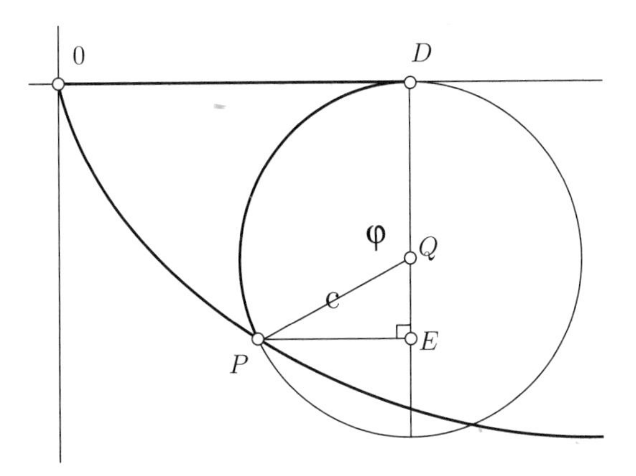

Figure 27.1. Cycloids, paths of points on rolling wheels, provide the fastest slides from the origin to lower points.

Solution of the brachistochrone problem. To get a formula for the sliding time T, we'll use conservation of energy. At a point along its descent from the origin to a point (a, b) with $a \geq 0$, $b > 0$, the puck has lost potential energy mgy, where m is its mass and g is the gravitational constant. This energy has been converted into kinetic energy:

$$\frac{1}{2}mv^2 = mgy.$$

Therefore,

$$\frac{ds}{dt} = v = \sqrt{2gy},$$

$$dt = \frac{ds}{\sqrt{2gy}} = \frac{\sqrt{1+y'^2}}{\sqrt{2gy}}dx,$$

and the elapsed time T satisfies

$$T = \int_0^a \sqrt{\frac{1+y'^2}{2gy}}dx.$$

Assuming for physical reasons that there is a smooth fastest path $y(x)$, it will satisfy Euler's equation. Since the integrand F does not depend explicitly on x, we have the first integral 27.1(1):

$$F - \frac{\partial F}{\partial y'} y' = C_1,$$

$$\frac{1}{\sqrt{2g}}(1 + y'^2)^{1/2} y^{-1/2} - \frac{1}{\sqrt{2g}} \frac{1}{2}(1 + y'^2)^{-1/2} \cdot 2y' y^{-1/2} \cdot 2y' = C_1.$$

Multiplying by $\sqrt{2g}(1 + y'^2)^{1/2} y^{1/2}$ yields

$$1 + y'^2 - y'^2 = C_2(1 + y'^2)^{1/2} y^{1/2},$$

$$y'^2 = \frac{2c}{y} - 1 = \frac{2c - y}{y},$$

$$\sqrt{\frac{y}{2c - y}}\, dy = dx,$$

where c is some new constant and the factor of 2 is chosen for future convenience. To integrate the left-hand side, you'd like to make the numerator and denominator of the fraction both squares. Use the substitution $y = 2c \sin^2 \theta$, $dy = 4c \sin^2 \theta \cos \theta \, d\theta$, to obtain

$$dx = 4c \sin^2 \theta \, d\theta = 2c(1 - \cos 2\theta) d\theta,$$

by the standard trigonometric identity for $\sin^2 \theta$. Integration yields

$$x = 2c \left(\theta - \frac{1}{2} \sin 2\theta \right) + C_3.$$

Since initially $x = 0$, $y = 0$, and hence $\theta = 0$, $C_3 = 0$. Therefore, if $\varphi = 2\theta$,

$$x = c(\varphi - \sin \varphi)$$

while

$$y = 2c \sin^2 \theta = c(1 - \cos \varphi),$$

the asserted cycloid.

Exercises 27

1. What is the fastest time to slide from the origin down to $(\pi/2 - 1, 1)$? How long would it take along a straight line? (Give answer in terms of gravitational acceleration g.)

2. Show that the fastest path goes below the final destination and comes back up if and only if $a/b > \pi/2$.

3. Find an interesting fact about the Bernoullis in a book or on the web.

4. What shape path would provide fastest descent if the puck were given an initial push? Would it be closer to or further from the straight line?

Chapter 28

Geodesics and Great Circles

In a surface like a sphere or a paraboloid, the shortest path between two points is smooth and has the property that the rate of change of its length under smooth variations vanishes. Such a path is called a *geodesic*. In other words, geodesics are smooth solutions to Euler's equation for the integral formula for length. All shortest paths are geodesics, but a geodesic need not be a shortest path; it could be just a local minimum or worse.

For example, on the sphere, the equator is a geodesic. Relatively short pieces of the equator are shortest paths, but if you go more than halfway around, it would be shorter to go in the other direction.

The equator is an example of a great circle, a largest possible circle on the sphere, the intersection of the sphere with a plane through its center. All great circles are geodesics. Moreover, all geodesics are great circles. The shortest path between New York and Madrid heads not due east along a circle of latitude, but at first a bit north of east along a great circle, as in Figure 28.1. This partly explains why flights to Europe go so far north on the way.

Theorem 28.2 will prove that geodesics are great circles, after Lemma 28.1 derives the equation for a great circle on the sphere in spherical coordinates φ and θ, where $0 \leq \theta < 2\pi$ is the usual planar angle and $0 \leq \varphi \leq \pi$ is the angle with the z-axis.

28.1. Lemma. *The equation of a great circle on the sphere is given by:*

$$\cot \varphi = a \sin(\theta + c),$$

except for vertical great circles ("circles of longitude") $\theta = c$.

Figure 28.1. The shortest path from New York to Madrid follows not a circle of latitude (below) but a great circle (above).

Proof. A nonvertical great circle is the intersection of the sphere with a plane $z = a_1x + a_2y$. In spherical coordinates,

$$\cos\varphi = a_1 \sin\varphi \cos\theta + a_2 \sin\varphi \sin\theta,$$
$$\cot\varphi = a_1 \cos\theta + a_2 \sin\theta.$$

Write the vector (a_2, a_1) in polar coordinates,

$$(a_2, a_1) = a(\cos c, \sin c),$$

to obtain

$$\cot\varphi = a(\sin c \cos\theta + \cos c \sin\theta) = a\sin(\theta + c). \qquad \square$$

28.2. Theorem. *Geodesics on a sphere are pieces of great circles.*

Proof. A nonvertical piece of geodesic away from the poles may be written as $u = f(\theta)$, with u defined as $\cot\varphi$. To compute the length of such a curve, recall from calculus that length in spherical coordinates is given by

$$ds^2 = d\varphi^2 + \sin^2\varphi \, d\theta^2. \tag{2}$$

Since $u = \cot\varphi$,

$$du = -\csc^2\varphi\, d\varphi = -(1+u^2)d\varphi,$$

$$d\varphi^2 = \left(\frac{du}{1+u^2}\right)^2 = \left(\frac{u'}{1+u^2}\right)^2 d\theta^2,$$

$$\sin^2\varphi\, d\theta^2 = \frac{d\theta^2}{\csc^2\varphi} = \frac{d\theta^2}{1+u^2},$$

Therefore by (1),

$$ds^2 = \left(\left(\frac{u'}{1+u^2}\right)^2 + \frac{1}{1+u^2}\right) d\theta^2$$

$$ds = (1+u^2)^{-1}(u'^2+1+u^2)^{1/2} d\theta.$$

Consequently, length L is given by the integral:

$$L = \int_{\theta_1}^{\theta_2} (1+u^2)^{-1}(u'^2+1+u^2)^{1/2} d\theta.$$

Since the integrand F does not mention θ explicitly, solutions have first integral 27.1(1):

$$F - u'\frac{\partial F}{\partial u'} = C,$$

or

$$(1+u^2)^{-1}(u'^2+1+u^2)^{1/2} - u'(1+u^2)^{-1}\cdot\frac{1}{2}(u'^2+1+u^2)^{-1/2}\cdot 2u' = C.$$

Multiplying both sides by $(1+u^2)(u'^2+1+u^2)^{1/2}$ yields

$$u'^2+1+u^2-u'^2 = C(1+u^2)(u'^2+1+u^2)^{1/2},$$

$$1 = C(u'^2+1+u^2)^{1/2},$$

$$u' = \pm\sqrt{a^2-u^2},$$

$$\frac{du}{\sqrt{a^2-u^2}} = \pm d\theta,$$

$$\sin^{-1}\frac{u}{a} = \pm(\theta+c),$$

$$u = a\sin(\theta+c),$$

after absorbing the $\pm$ sign into the constant a. Therefore every nonvertical geodesic is a piece of a great circle. The steps are reversible, so every great circle is a geodesic.

If a geodesic ever goes vertical or through the poles, rotate it tangent to the equator, infer that it now coincides with the equator, and conclude that it must have been a vertical great circle to start with. □

Remark. For the round sphere, the curvature vector of a geodesic is always normal to the surface. This turns out to be true for every smooth surface, and can be used as an alternative definition of a geodesic.

Exercises 28

1. Use the calculus of variations to show that geodesics in $\mathbb{R}^2$ are straight lines.

2. Describe geodesics on a unit cylinder. How many geodesics are there between two points?

3. Assuming that New York and Madrid are both at latitude 41 degrees north and at longitude 74 and 3 degrees west, respectively, calculate the distance along a geodesic and the distance along a circle of latitude. Assume that the earth is a round sphere of radius 4000 miles.

4. On the round sphere, at every point, there is a unique geodesic in every direction. Do you think that this is true on every smooth surface?
Using just this fact, use a symmetry argument to prove that geodesics on a round sphere are great circles.

Chapter 29

Variational Notation, Higher Order Equations

29.1. Variational Notation. In calculus one often thinks of the derivative df/dx as the ratio of an infinitesimal change in $f(x)$ to the infinitesimal change in x which causes it. Similarly, in the calculus of variations one often uses δI to denote an infinitesimal change in some cost $C = \int F$ due to some infinitesimal change $\delta y = \eta \, d\varepsilon$ in the unknown function $y(x)$. This is really just useful shorthand for something perfectly rigorous. For example, in such variational notation, the derivation of Euler's Equation 26.5

$$\frac{\partial F}{\partial y} = \frac{d}{dx}\frac{\partial F}{\partial y'}$$

for a minimum of the cost

$$C = \int_a^b F(x, y, y') \, dx$$

takes the following form:

Derivation of Euler's Equation 26.5 in variational notation. At a minimum, for any smooth infinitesimal variation δy vanishing at the endpoints,

$$\begin{aligned} 0 = \delta C &= \int_a^b \delta F dx \\ &= \int_a^b \frac{\partial F}{\partial y}\delta y + \frac{\partial F}{\partial y'}\delta y'\,dx \quad \text{(by the Chain Rule)} \\ &= \int_a^b \frac{\partial F}{\partial y}\delta y - \frac{d}{dx}\frac{\partial F}{\partial y'}\delta y\,dx \quad \text{(integration by parts)} \\ &= \int_a^b \left(\frac{\partial F}{\partial y} - \frac{d}{dx}\frac{\partial F}{\partial y'}\right)\delta y\,dx. \end{aligned}$$

Since this holds for all smooth infinitesimal changes vanishing at the endpoints, it follows from the Fundamental Lemma of the Calculus of Variations 26.4 that

$$\frac{\partial F}{\partial y} - \frac{d}{dx}\frac{\partial F}{\partial y'} = 0$$

as desired.

More generally, for a cost integral $C = \int F$, F may depend on higher derivatives of y, or y may depend on several variables instead of just x. In such cases, the corresponding Euler equation becomes a bit more complicated. Here are two representative examples.

29.2. Second order. *A minimum $y = f(x)$, among all C^3 functions, of the cost*

$$\int_a^b F(x, y, y', y'')dx,$$

given $f(a)$ and $f(b)$, satisfies the Euler equation

$$\frac{\partial F}{\partial y} - \frac{d}{dx}\frac{\partial F}{\partial y'} + \frac{d^2}{dx^2}\frac{\partial F}{\partial y''} = 0.$$

(Here we assume that F has three continuous partial derivatives.)

Proof. Exercise 1. □

When instead of x there are two independent variables, it is customary to call them x and y, and use perhaps u instead of y for the unknown function. The partial derivatives of $u(x, y)$ are denoted by u_x and u_y. The region of integration is no longer an interval $[a, b]$ but a bounded domain D in the x-y plane.

29.3. Two independent variables. *A minimum $u = f(x, y)$, among all C^2 functions, of the cost*

$$\int_D F(x, y, u, u_x, u_y)\, dx\, dy,$$

with given values on the boundary of D, satisfies the Euler equation

$$\frac{\partial F}{\partial u} - \frac{\partial}{\partial x}\frac{\partial F}{\partial u_x} - \frac{\partial}{\partial y}\frac{\partial F}{\partial u_y} = 0.$$

Proof. Exercise 3. □

29.4. Gradient and divergence. On a smooth planar domain D, the symbol $\nabla = \left(\frac{\partial}{\partial x}, \frac{\partial}{\partial y}\right)$, called "del," is used to define the vector gradient of a C^1 function $f(x, y)$ and the scalar divergence of a C^1 vectorfield $\boldsymbol{V}(x, y) = (g(x, y), h(x, y))$ as follows:

$$\operatorname{grad} f = \nabla f = \left(\frac{\partial f}{\partial x}, \frac{\partial f}{\partial y}\right), \quad \operatorname{div} \boldsymbol{V} = \nabla \bullet (g, h) = \frac{\partial g}{\partial x} + \frac{\partial h}{\partial y}.$$

The gradient of f is a vector encoding all the information about f's derivatives. The divergence of a vectorfield makes its dramatic appearance in Theorem 29.5.

29.5. Gauss's Divergence Theorem. *Let $\boldsymbol{V}$ be a C^1 vectorfield on a piecewise C^1 region R in the plane. Then*

$$\int_{\partial R} \boldsymbol{V} \bullet \boldsymbol{n} = \int_R \operatorname{div} \boldsymbol{V}.$$

If $\boldsymbol{V}$ represents the velocity of a flow, then the left-hand side represents the net flow out through the boundary of R. Hence $\operatorname{div} \boldsymbol{V}$ must represent the amount of stuff being created, the sources minus the sinks.

Proof. Exercise 4 proves the case of a square, the only case we will need. We omit the proof of the general case. □

29.6. Higher dimensions. Everything in this chapter extends immediately from $\mathbb{R}^2$ to $\mathbb{R}^n$.

Exercises 29

1. Use variational notation to derive 29.2.

2. Verify integration by parts for a square $D = [a, b] \times [c, d]$ in the plane: for C^1 functions $f(x, y)$, $\eta(x, y)$, with η vanishing on the boundary of D,

$$\int_D f \frac{\partial \eta}{\partial x} = -\int_D \frac{\partial f}{\partial x} \eta.$$

Recall that by Fubini's Theorem 17.3,

$$\int_D g(x, y)\, dA = \int_c^d \int_a^b g(x, y)\, dx\, dy = \int_a^b \int_c^d g(x, y)\, dy\, dx.$$

(Such integration by parts holds for any bounded domain D.)

3. Use variational notation and Exercise 2 to derive 29.3. Since it suffices to derive the Euler equation at every point, it suffices to take D to be a small square about the point.

4. From Exercise 2, verify for a square the following vector form of integration by parts for a C^1 function $f(x, y)$ and C^1 vectorfield $\boldsymbol{V}(x, y)$, one of which vanishes on the boundary:

$$\int_D \nabla f \bullet \boldsymbol{V} = -\int_D f \operatorname{div} \boldsymbol{V}.$$

5. Show that $\operatorname{div}(f\boldsymbol{V}) = \nabla f \bullet \boldsymbol{V} + f \operatorname{div} \boldsymbol{V}$.

6. Use Exercise 5 and the Divergence Theorem 29.5 to give another proof of Exercise 4.

7. Prove the Divergence Theorem 29.5 for a square $R = [a, b] \times [c, d]$ in the plane. If $\boldsymbol{V} = (u, v)$, it suffices to prove it for each component, e.g., for $(u, 0)$.

8. In vector notation, the length of a curve $\boldsymbol{x}(t)$ in $\mathbb{R}^n$ is given by

$$L = \int_{t_1}^{t_2} (\boldsymbol{x} \bullet \boldsymbol{x})^{1/2} dt.$$

Show that for a smooth curve parameterized by arclength, so that $\dot{\boldsymbol{x}} = d\boldsymbol{x}/dt$ is the unit tangent vector $\boldsymbol{T}$, $\delta L = 0$ means that $\frac{d}{dt}\boldsymbol{T} = 0$, i.e., that $\boldsymbol{x}$ is a straight line.

Chapter 30

Harmonic Functions

30.1. Definition. A function $f(x, y, z)$ is *harmonic* if the Laplacian Δf vanishes:

$$\Delta f = f_{xx} + f_{yy} + f_{zz} = 0, \tag{1}$$

where subscripts denote partial derivatives. Note that

$$\Delta f = \operatorname{div} \operatorname{grad} f.$$

The Laplacian is a three-dimensional sort of second derivative. Just as the second derivative of a function $f(x)$ of one variable tells whether nearby values are greater or smaller than the value at a point, the Laplacian of a function $f(x, y, z)$ of three variables tells how average nearby values differ from the value at a point. In particular, the vanishing of the Laplacian, Laplace's equation (1), means that the value at a point equals the average of nearby values. Indeed, at a point on the boundary of a small ball B about a point p, the rate of change of f as you move away from p is $(\operatorname{grad} f) \bullet \boldsymbol{n}$. By the Divergence Theorem 29.5, its average value is proportional to

$$\int_{\partial B} (\operatorname{grad} f) \bullet \boldsymbol{n} = \int_B \operatorname{div} \operatorname{grad} f = \int_B \Delta f \approx 0,$$

so that on the average, f is not changing as you move away from p.

There are many important examples of harmonic functions; we'll consider two:

(1) A steady-state (time independent) temperature distribution, since the only way that the temperature will not change in time is if the temperature at a point equals the average of the nearby temperatures.

(2) Gravitational potential energy φ in empty space (e.g., not inside the earth). It is perhaps surprising that gravitational potential energy has a value at every point equal to the average of the values at nearby points. To see that φ is indeed harmonic, consider the gravitational forcefield $F = -\operatorname{grad}\varphi$ (potential energy increases as you do work *against* gravity). Since F has no sources in empty space,

$$0 = \operatorname{div} F = -\operatorname{div}\operatorname{grad}\varphi = -\Delta\varphi.$$

More generally (e.g., inside the earth), the Laplacian of gravitational potential energy is proportional to the mass density.

One way of measuring the amount a function changes is the *Dirichlet energy*

$$E = \int |\nabla f|^2.$$

The following theorem shows that functions minimizing Dirichlet energy are harmonic. Physics likes to be able to solve its differential equations (such as Laplace's equation) just by minimizing some energy.

30.2. Theorem. *A C^2 minimizer of the Dirichlet energy (for given boundary values) is harmonic.*

Proof 1. Exercise 3 shows that the Euler equation reduces to Laplace's equation. □

Proof 2. Exercise 4 directly derives Laplace's equation using the calculus of variations. □

Proof 3. For good measure, we give an elegant vector proof. For a minimizer,

$$\begin{aligned} 0 = \delta E = \delta \int |\nabla f|^2 &= \delta \int \nabla f \bullet \nabla f \\ &= \int 2\nabla f \bullet \delta\nabla f \quad \text{(by Exercise 3)} \\ &= -2\int \Delta f \bullet \delta f \quad \text{(by Exercise 29.4)}, \end{aligned}$$

so that $\Delta f = 0$. □

Exercises 30

1. Near the surface of the earth, gravitational potential energy

$$\varphi = cz.$$

Verify that φ is harmonic.

2. In the solar system, the gravitational potential energy φ of a mass m due to the sun (of mass M) satisfies

$$\varphi = -GMm/r = -c(x^2 + y^2 + z^2)^{-1/2}.$$

a. Verify directly that φ is harmonic.

b. Verify that φ is harmonic in two steps. First check that

$$F = -\operatorname{grad}\varphi = -\frac{GMm}{r^2}\boldsymbol{r},$$

where $\boldsymbol{r}$ is a unit radial vector (Newton's inverse square law of gravity). Then check that $\operatorname{div} F = 0$.

3. Prove Theorem 30.2 by showing that the Euler equation reduces to Laplace's equation.

4. Derive Theorem 30.2 directly.

5. Show that for two C^1 vector functions $\boldsymbol{V}, \boldsymbol{\eta}$,

$$\frac{d}{d\varepsilon}((\boldsymbol{V} + \varepsilon\boldsymbol{\eta}) \bullet (\boldsymbol{V} + \varepsilon\boldsymbol{\eta}))|_{\varepsilon=0} = 2\boldsymbol{V} \bullet \boldsymbol{\eta}.$$

In variational notation this means that

$$\delta(\boldsymbol{V} \bullet \boldsymbol{V}) = 2\boldsymbol{V} \bullet \delta\boldsymbol{V},$$

the expected product rule.

6. Prove that the sums and multiples of harmonic functions are harmonic.

Chapter 31

Minimal Surfaces

Many surfaces in nature tend to minimize area. The best examples are soap films, as pictured in Figure 31.1. If the graph of a function f minimizes area A, then f satisfies an important partial differential equation, the minimal surface equation, equivalent to $\delta A = 0$. Such surfaces are called minimal surfaces. Some famous minimal surfaces are pictured in Figure 31.2.

Figure 31.1. Soap films tend to minimize area for given boundary. "Garden of minimal surfaces," models (and photograph) © J. C. C. and C. D. Nitsche, from Nitsche's *Lectures on Minimal Surfaces*, Cambridge University Press, 1989, used by permission, all rights reserved.

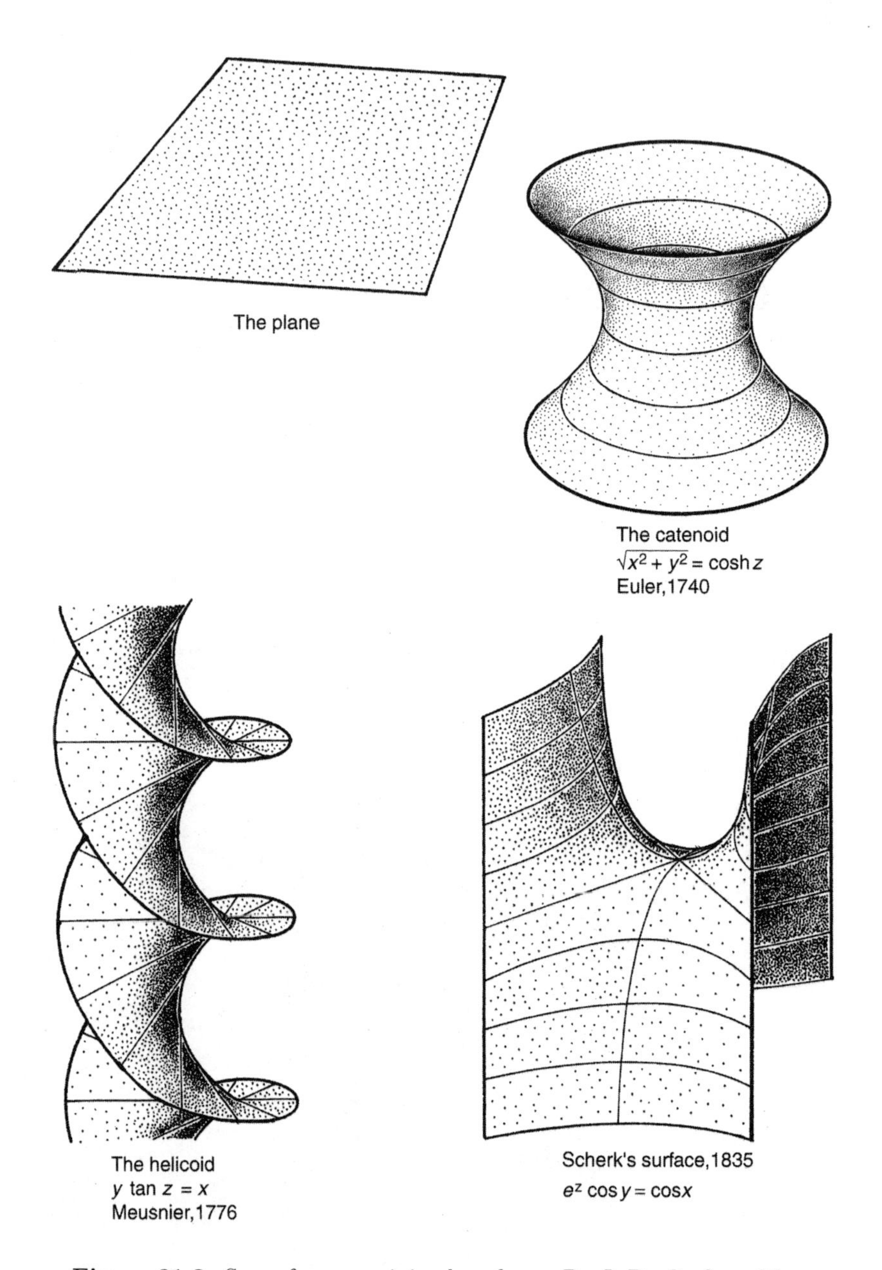

Figure 31.2. Some famous minimal surfaces. By J. Bredt, from Morgan's *Geometric Measure Theory*, © Frank Morgan.

31.1. Theorem. *Let f be a real-valued function on a planar domain D such that the graph of f is a local minimum for area among graphs of C^2 functions with given boundary. Then f satisfies the minimal surface equation*:

$$(1+f_y^2)f_{xx} - 2f_x f_y f_{xy} + (1+f_x^2)f_{yy} = 0. \tag{1}$$

This is a much harder equation to deal with than earlier equations such as Laplace's equation because derivatives are squared and multiplied: the equation is *nonlinear.* Sums $f + g$ and multiples cf of solutions need not be solutions because of the interference you get when you plug in $f + g$ or cf.

Notice that for nearly flat surfaces, surfaces for which f_x and f_y are small, the minimal surface equation is approximately the same as Laplace's equation, and the value at a point is approximately the average of nearby values. As f_x and f_y increase, the behavior becomes much more complicated. Such nonlinearity sometimes causes stretched soap films to break apart, or oscillating soap films to give birth to soap bubbles.

Proof. The formula from calculus for area A is given by

$$A = \int_D \sqrt{1 + f_x^2 + f_y^2}.$$

Exercise 3 verifies that (1) is the associated Euler equation. □

Exercises 31

1. Verify that the helicoid of Figure 31.2 is a minimal surface.

Hint: Permute variables to $z = f(x, y) = x \tan y$ before plugging into the minimal surface equation 31.1(1).

2. Verify that Scherk's surface $\left\{z = \log \frac{\cos x}{\cos y} : \ -\pi/2 < x, y < \pi/2\right\}$ is a minimal surface.

Hint: $z = \log \cos x - \log \cos y$.

3. Verify that the minimal surface equation 31.1(1) is the Euler equation for the area functional.

4. For a real-valued function on a domain in $\mathbb{R}^n$, the area A is given by

$$A = \int_D \sqrt{1 + (\nabla f)^2},$$

where $(\nabla f)^2$ means $(\nabla f) \bullet (\nabla f)$. Derive from scratch the general minimal surface equation

$$\operatorname{div} \frac{\nabla f}{\sqrt{1 + (\nabla f)^2}} = 0.$$

5. Verify that the equations of Exercise 4 reduce to the planar formulas for $n = 2$.

6. Prove that the catenoids $\sqrt{x^2 + y^2} = \frac{1}{a} \cosh az$ are the only smooth, minimal surfaces of revolution in $\mathbb{R}^3$.

Chapter 32

Hamilton's Action and Lagrange's Equations

Hamilton's "Principle of Least Action," together with the calculus of variations, provides a very general and powerful method for solving problems in physics and mechanics, such as Exercise 4. The principle of least action actually goes back to Maupertuis and Euler.

A physical particle may have some kinetic energy $T = \frac{1}{2}mv^2$ and some potential energy V. The total energy is the sum $T + V$. Hamilton proposed as a fundamental principle that Nature tends to minimize the average over time of the *difference* $T - V$, called the *action* or *Lagrangian.* From this perspective, a falling object speeds up as it falls because it wants to spend less time low where T is big and V is small, and more time high where T is small and V is big. You might argue that it would never want to come down at all, but the starting and ending positions and times are given. So given that it ends up on the ground at a certain time, it prefers to speed up as it descends. But it does not want to speed up too much, or the term containing the *square* of the velocity will make the action too big. The precise balance to minimize the action, the associated Euler equation, turns out to be

$$F = ma,$$

thus providing an amazing derivation of Newton's fundamental law of mechanics.

32.1. Theorem. *A particle (on a smooth path) minimizing action satisfies*

$$F = ma.$$

Proof. The proof uses little more than the fact that force is the negative gradient of the potential: $F = -\nabla V$. It also uses the chain rule: $\delta V(\boldsymbol{x}) = \nabla V \bullet \delta\boldsymbol{x}$, perhaps more familiar to you as

$$\frac{dV}{dt} = \frac{\partial V}{\partial x}\frac{dx}{dt} + \frac{\partial V}{\partial y}\frac{dy}{dt} + \frac{\partial V}{\partial z}\frac{dz}{dt}.$$

Let $\boldsymbol{x}$ denote position, so that $\dot{\boldsymbol{x}}$ denotes velocity and $\ddot{\boldsymbol{x}}$ denotes acceleration. Then at a minimum,

$$\begin{aligned} 0 &= \delta \int T - V\,dt \\ &= \delta \int \frac{1}{2}m\dot{\boldsymbol{x}}^2 - V\,dt \\ &= \int m\dot{\boldsymbol{x}} \bullet \delta\dot{\boldsymbol{x}} - \nabla V \bullet \delta\boldsymbol{x}\,dt \quad \text{(by Leibniz and the chain rule)} \\ &= \int -m\ddot{\boldsymbol{x}} \bullet \delta\boldsymbol{x} + F \bullet \delta\boldsymbol{x}\,dt \quad \text{(by integration by parts),} \end{aligned}$$

so that $0 = -m\ddot{\boldsymbol{x}} + F$, i.e., $F = ma$. □

32.2. Lagrange's equations. Now we consider a system of n particles, with "positions" $q_1, \ldots, q_n$ and "velocities" $\dot{q}_1, \ldots, \dot{q}_n$. Typically each q_i would be a vector in $\mathbb{R}^3$ and hence $\dot{q}_i$ would be the associated velocity vector, but q_i could be another parameter, such as an angle. One assumes that the kinetic energy T is a function of the q_i and $\dot{q}_i$, while the potential energy V depends on the q_i alone. Then the Euler equation for least action is

$$\frac{d}{dt}\frac{\partial T}{\partial \dot{q}_i} - \frac{\partial T}{\partial q_i} + \frac{\partial V}{\partial q_i} = 0.$$

32.3. The pendulum. For example, consider a pendulum of length 1 with bob of mass m, with position q_1 given by angle θ with the vertical. The kinetic energy $T = \frac{1}{2}m\dot{\theta}^2$, while the potential energy $V = mgh = mg(1 - \cos\theta)$. Hence, Lagrange's equation yields

$$\ddot{\theta} + g\sin\theta = 0, \tag{1}$$

the standard differential equation for the pendulum.

Exercises 32

1. Verify that Lagrange's equations are just the Euler equation for least action.

2. Verify that Lagrange's equations yield the pendulum differential equation 32.3(1).

3. Give the more familiar derivation of 32.3(1) from Newton's $F = ma$.

4. Show that Lagrange's equation gives the standard differential equation $\ddot{h} = -g$ for the height $q_1 = h$ of a particle falling with gravitational acceleration g.

5. A frictionless puck (see figure next page) follows a circle of radius a about a small hole in a table, balanced by a bob dangling from a string connected to the puck through the hole. Puck and bob both have unit mass. At time 0, the bob is pulled downward a bit and released, while the puck continues to revolve. What happens? Will the puck spiral inward to the hole?

a. Show that for equilibrium, the angular velocity ω of the puck must satisfy $\omega^2 = g/a$, where g is gravitational acceleration.

b. Show that the Lagrangian (action) for puck and bob in terms of the angle θ for the puck and the displacement x of the bob below its initial position is

$$T - V = \dot{x}^2 + \frac{1}{2}(a - x)^2\dot{\theta}^2 + gx.$$

Hint: The square of the velocity of the puck is the sum of the squares of its tangential and inward components: $(a - x)^2\dot{\theta}^2 + \dot{x}^2$.

c. Show that Lagrange's equations become

$$0 = \frac{d}{dt}(2\dot{x}) + (a - x)\dot{\theta}^2 - g,$$

$$0 = \frac{d}{dt}((a - x)^2\dot{\theta}).$$

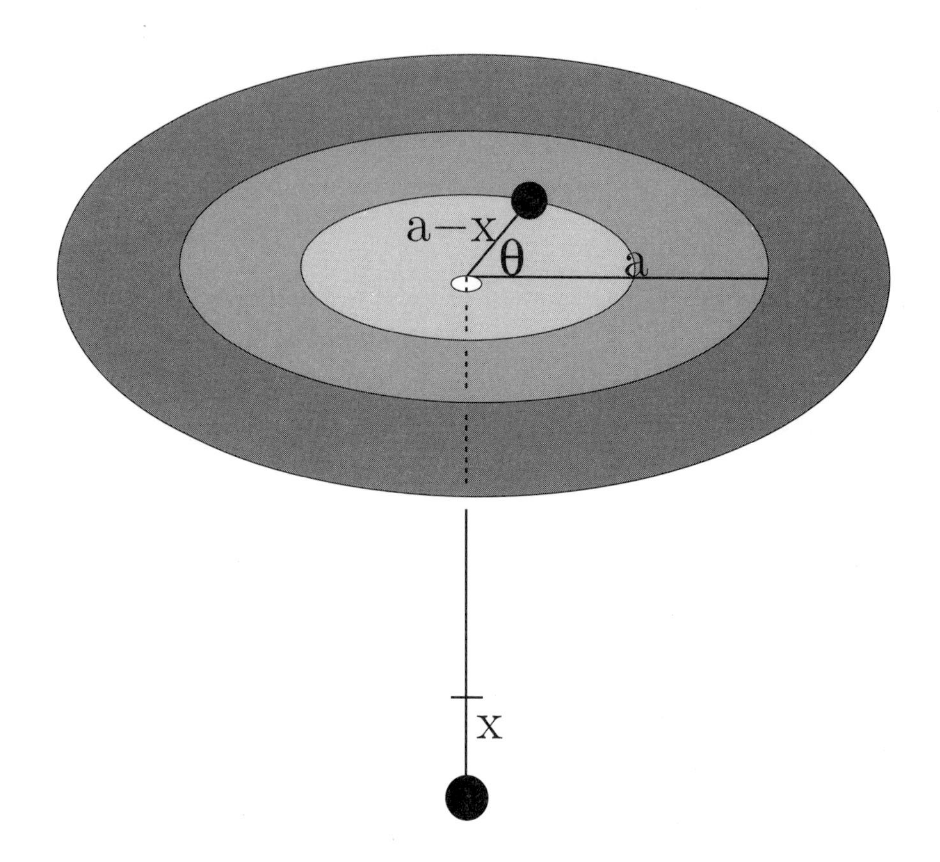

d. Obtain this differential equation for x alone:

$$2\ddot{x} + [(1 - x/a)^{-3} - 1]g = 0.$$

Hint: From the second equation, $(a - x)^2\dot{\theta} = c$. To evaluate c, use the fact from part a that when $x = 0$, $\dot{\theta}^2 = g/a$. Now solve for $\dot{\theta}$ and plug into the first equation.

e. For x small, obtain the approximation

$$2\ddot{x} + 3gx/a = 0,$$

with solution $x = \cos\sqrt{3g/2a}\,t$. In other words, the bob performs small vertical oscillations of frequency $\sqrt{3g/2a}$; it does not spiral inward.

Chapter 33

Optimal Economic Strategies

33.1. Example. The Euler Wagon Company needs to produce B children's wagons in time T (perhaps 5000 in 100 days). Let $x(t)$ denote the inventory, the number of wagons on hand, with $x(0) = 0$ and $x(T) = B$. Then the production rate is $x'(t)$ wagons/day.

A higher than usual production rate can be costly: renting extra, more expensive machines, paying workers more expensive overtime, and paying a premium to get extra parts fast. For simplicity, we'll assume that the unit production cost is proportional to the production rate, $c_1 x'$ dollars/wagon, yielding a production cost per day of

$$(c_1 x' \text{ dollars/wagon})(x'(t) \text{ wagons/day}) = c_1 x'^2 \text{ dollars/day}.$$

In addition, we'll assume that the inventory cost per day for storing the wagons until they are needed is proportional to the number of wagons in storage: $c_2 x$ dollars/day. Our goal is to minimize the total cost C, the sum of production and inventory costs, given by

$$C = \int_0^T (c_1 x'^2 + c_2 x)\, dt. \tag{1}$$

Before solving, let's try to guess the answer. If there were no inventory cost, it would probably be best to produce at an even, constant rate: $x = \frac{B}{T} t$. With the inventory cost, it is probably better to start slower and increase the rate of production; perhaps x might grow quadratically.

Solution. Assuming a smooth minimum $x(t)$, by Euler's equation,

$$0 = \frac{d}{dt}(2c_1x') - c_2 = 2c_1x'' - c_2,$$
$$x'' = c_2/2c_1,$$

x'' is constant and x is indeed quadratic. Integrating twice and plugging in the initial conditions to solve for the two constants of integration yield

$$x = B\frac{t}{T} - \frac{c_2}{4c_1}t(T-t), \tag{2}$$

as in Figure 33.1.

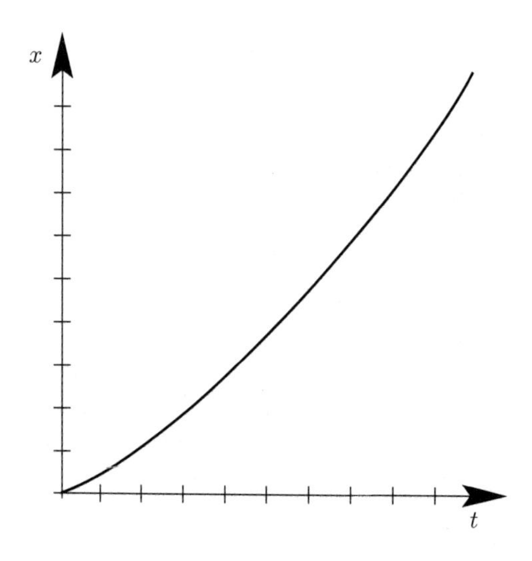

Figure 33.1. The optimal inventory x grows quadratically.

33.2. Remark. What is the significance of negative inventory x as in Figure 33.2a? One interpretation is that storage is so expensive relative to our production costs that we would like not only to hold off on production but make a deal with our friendly competitors to meet their production needs later if they pay us what they save on storage costs. If no such opportunity is available, the best strategy is to start production at a time $t_1 > 0$ with $x'' = c_2/2c_1$ such that $x(t_1) = x'(t_1) = 0$, the only possible C^1 solution, as pictured in Figure 33.2b.

It follows from (2) that such a situation ($x'(0) < 0$) arises when

$$\frac{c_2T}{4c_1} > \frac{B}{T},$$

i.e., when the storage cost constant c_2 is large relative to the production constant c_1, or if the due date T is very distant. Exercise 2 shows that the ideal amount of time to wait to start production is

$$t_1 = T - \sqrt{4c_1 B/c_2}.$$

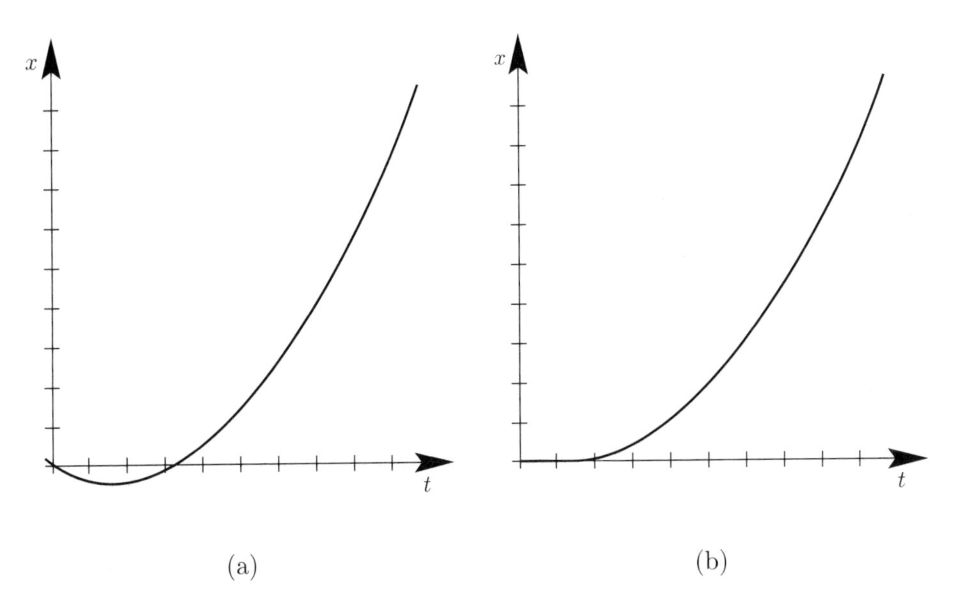

Figure 33.2. If the optimal solution $x(t)$ goes negative, the best practical solution starts later with zero slope and the same second derivative.

33.3. Discounting. There is another advantage to delaying production: in the meantime our money can be earning interest in the bank. After time t in the bank at interest rate r compounded continuously, our balance gets multiplied by e^{rt}. To take this into account, we should multiply future expenses by e^{-rt}. In Example 33.1, we should really minimize

$$C = \int_0^T e^{-rt}(c_1 x'^2 + c_2 x)\, dt. \tag{3}$$

Then the Euler equation yields

$$x'' = \frac{c_2}{2c_1} + rx',$$

so that production increases more rapidly as the due date approaches, as you would expect. See Exercise 3.

The assumption of a production cost per day of $g(x') = c_1 x'^2$ may not hold, but the following proposition shows that the same qualitative behavior still holds as long as the cost accelerates ($g'' > 0$).

33.4. Proposition. *Suppose that the production cost per day $g(x')$ satisfies:*

$$g(0) = 0, \quad g'' > 0.$$

Then for any smooth minimizer x with x and x' nonnegative,

$$x'' > 0,$$

i.e., the production rate x' is increasing.

Proof. Exercise 4. □

Exercises 33

1. Derive equation (2), starting with equation (1).

2. Finish the analysis of Remark 33.2.

3. Completely work the example of 33.3, finally obtaining the solution

$$x = \left(B + \frac{c_2}{2c_1 r}T\right)\frac{e^{rt} - 1}{e^{rT} - 1} - \frac{c_2}{2c_1 r}t.$$

Hint: To solve the Euler equation conclusion $x'' - rx' = c_2/2c_1$, multiply both sides by the "integrating factor" e^{-rt}, making the lefthand side $(e^{-rt}x')'$, and hence $e^{-rt}x' = (c_2/2c_1)t + a$.

4. Prove Proposition 33.4.
Hint: Show that $g''(x')x'' = rg'(x') + c_2$. Since g'' and hence g' are positive, x'' nust be positive.

Chapter 34

Utility of Consumption

How much of your earnings should you save? Suppose you know you have T years to live. The way to have the most money to spend is to invest everything to grow at some rate i and spend all your accrued millions just before you die. But the happiness or "utility" that the last dollar among millions will provide on that glorious future day is much less than its utility to you today when you are otherwise going hungry. The first dollars you spend give you more happiness than additional ones. Ten times as much spending does not make a day ten times as happy, maybe just twice as happy. For an amount of spending or *consumption* C, there is an associated *utility* $U(C)$ which grows more slowly as C increases, as in Figure 34.1, maybe like $\log C$ or $C^{1/2}$ or C^a for $0 < a < 1$. You want to maximize utility over your life:

$$\int_0^T U(C(t))\, dt.$$

We'll take your wages $v(t)$ as a given, "exogenously determined." Let $K(t)$ be your capital investments, growing at some rate i, yielding additional income iK. (If you borrow money, then K is negative, and the negative income iK represents your loan or mortgage payments.) Hence your total income is $v + iK$, which is divided somehow between your consumption C and adding to your capital:

$$iK(t) + v(t) = C(t) + K'(t). \tag{1}$$

Your starting capital $K(0)$ is given, and to maximize your lifetime utility you're planning to spend everything: $K(T) = 0$.

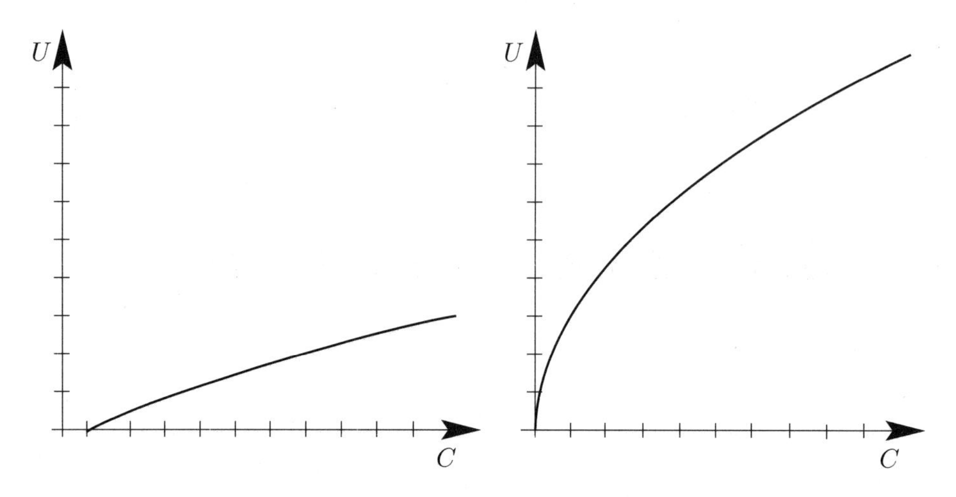

Figure 34.1. Utility $U(C)$ grows more slowly as consumption C increases, perhaps like $\log C$ or $C^{1/2}$.

Equation (1) determines C in terms of K, K', and t. Our primary unknown function is $K(t)$. For a smooth maximum, the Euler equation yields

$$\begin{aligned} 0 &= \frac{d}{dt}\frac{\partial U}{\partial K'} - \frac{\partial U}{\partial K} = \frac{d}{dt}\left(U'\frac{\partial C}{\partial K'}\right) - U'\frac{\partial C}{\partial K} \\ &= \frac{d}{dt}(U'(-1)) - U'i = -U''C' - U'i. \end{aligned} \tag{2}$$

Now assume that $U = \log C$, $U' = C^{-1}$, $U'' = -C^{-2}$. Then (2) becomes

$$\begin{aligned} 0 &= C^{-2}C' - C^{-1}i, \\ C' &= iC, \\ C &= C_0 e^{it}. \end{aligned} \tag{3}$$

A beautiful result: one should so arrange one's finances, so that spending grows at the same rate as investments.

34.1. Inflation. If there is an inflation rate r, then the utility of future consumption will be just $U(e^{-rt}C)$. Exercises 3 and 4 maximize utility with inflation.

Exercises 34

1. Show that in the case of no wages, the constant C_0 of equation (3) is equal to $K(0)/T$. In other words, if you have 20 years to live, spend 1/20 of your money this year.

Hint: Go back to (1) and solve for K.

2. Show that for $U(C) = C^a$ $(0 < a < 1)$, ideal consumption (3) becomes

$$C = C_0 \exp\left(\frac{i}{1-a}t\right).$$

The closer a is to 1, the faster consumption grows, approaching the limit scenario of spending everything at the last minute.

3. Show that with inflation, for $U = (e^{-rt}C)^a$, ideal consumption (3) becomes

$$C = C_0 \exp\left(\frac{i-ar}{1-a}t\right).$$

Note that with inflation, consumption grows more slowly and hence starts larger: there is more incentive to spend money early on while it's worth more.

Hint: Since U now depends on t as well as on C, you need to redo equation (2).

4. Continuing Exercise 3, assuming no wages $(v = 0)$, show that

$$C_0 = \frac{(b-i)K_0}{e^{(b-i)T} - 1},$$

where $b = \frac{i-ar}{1-a}$.

Hint: By equation (1),

$$K' - iK = C(t) = -C_0 e^{bt}.$$

If you've taken a course in differential equations, you know that the way to integrate an equation like this ("first-order linear") is to multiply both sides by the "integrating factor" e^{-it}. Then integration yields

$$e^{-it}K = -\frac{C_0}{b-i}e^{(b-i)t} + C_1.$$

Now plug in $K(0) = K_0$ and $K(T) = 0$ and solve for C_0.

Chapter 35

Riemannian Geometry

Riemannian geometry permits one to measure distance inside a surface by generalizing the usual formula for arclength. It turns out that that is all you need to do geometry.

35.1. Arclength. Recall from calculus that an element of planar arclength ds satisfies

$$ds^2 = dx^2 + dy^2 \tag{1}$$

(an infinitesimal Pythagorean theorem). For example, for a curve $y = f(x)$, $a \leq x \leq b$, the length L is given by

$$L = \int_a^b ds = \int_a^b \sqrt{1 + y'^2}\, dx.$$

On a circle of radius a, it is more convenient to put everything in terms of θ, and $ds = a\, d\theta$. On a sphere of radius a in spherical coordinates φ and θ,

$$ds^2 = a^2\, d\varphi^2 + a^2 \sin^2 \varphi\, d\theta^2. \tag{2}$$

More generally, on any surface with any coordinates u, v, ds^2 is some linear combination of du^2, dv^2, and, if the coordinates are not orthogonal, $du\, dv$. The coefficients are some functions which we'll call g_{ij}.

$$\begin{aligned} ds^2 &= g_{11}\, du^2 + g_{12}\, du\, dv + g_{21}\, dv\, du + g_{22}\, dv^2 = \\ &= g_{11}\, du^2 + 2g_{12}\, du\, dv + g_{22}\, dv^2. \end{aligned} \tag{3}$$

In example (1), $g_{11} = 1$, $g_{22} = 1$, and $g_{12} = g_{21} = 0$. In example (2), $g_{11} = a^2$ and $g_{22} = a^2 \sin^2 \varphi$.

Along a curve, the coordinates u and v are functions of time t, with derivatives $\dot{u}$ and $\dot{v}$. Putting everything in terms of t yields a formula for arclength L:

$$L = \int_{t_1}^{t_2} \sqrt{g_{11}\dot{u}^2 + 2g_{12}\dot{u}\dot{v} + g_{22}\dot{v}^2}\, ds. \tag{4}$$

A *Riemannian surface* is just a surface with a formula like this for computing length, called a Riemannian metric, determined by the g_{ij}'s. The notion generalizes immediately to higher dimensional Riemannian manifolds.

35.2. Geodesics. By definition, a geodesic satisfies $\delta L = 0$. As we saw in Chapter 28, geodesics in the plane are straight lines and geodesics on the sphere are great circles. In a general Riemannian surface, the geodesics can be much more complicated. A lot is known about geodesics. For example, in a smooth Riemannian manifold, there is a unique geodesic from every point in every direction.

Assuming to keep formulas simpler that our curve $u(t)$, $v(t)$ is parameterized by a multiple of arclength (constant speed), setting $\delta L = 0$ eventually leads to a formidable system of two differential equations for u and v:

$$\begin{aligned} 0 &= \ddot{u} + \Gamma^1_{11}\dot{u}^2 + 2\Gamma^1_{12}\dot{u}\dot{v} + \Gamma^1_{22}\dot{v}^2, \\ 0 &= \ddot{v} + \Gamma^2_{11}\dot{u}^2 + 2\Gamma^2_{12}\dot{u}\dot{v} + \Gamma^2_{22}\dot{v}^2, \end{aligned} \tag{1}$$

where the *Christoffel symbols* Γ^i_{jk} are given by the formula

$$\Gamma^i_{jk} = \frac{1}{2}\sum_{\ell} g^{i\ell}(g_{\ell j,k} + g_{\ell k,j} - g_{jk,\ell}), \tag{2}$$

where a subscript after a comma denotes partial differentiation with respect to u or v, according to whether the subscript is 1 or 2, and the $g^{i\ell}$are the components of the inverse matrix of $g_{i\ell}$. Note that $\Gamma^i_{kj} = \Gamma^i_{jk}$.

For example, for the easiest, Euclidean case

$$ds^2 = dx^2 + dy^2,$$

$g_{11} = g_{22} = 1$, $g_{12} = 0$, $[g_{ij}]$ is the identity matrix, $[g^{ij}]$ is also the identity matrix, $g^{11} = g^{22} = 1$, $g^{12} = 0$, the Γ^i_{jk} are all 0, and the equations for geodesics (1) become:

$$\begin{aligned} 0 &= \ddot{u}, \\ 0 &= \ddot{v}. \end{aligned}$$

The solutions are straight lines, as geodesics in the Euclidean plane should be.

More generally, on an n-dimensional surface with n coordinate functions $u_1, u_2, \ldots, u_n$, there are n geodesic equations: for each $1 \leq i \leq n$,

$$0 = \ddot{u}_i + \sum \Gamma^i_{jk} \dot{u}_j \dot{u}_k. \tag{3}$$

Since $\Gamma^i_{kj} = \Gamma^i_{jk}$, you really get the terms with $j \neq k$ twice, as in equation (1) above.

Exercises 35

1. Use $ds = a\, d\theta$ to compute that the length of a circle of radius a is $2\pi a$.

2. Use 35.1(2) to compute the lengths of circles of latitude and longitude on a sphere of radius a.

3. Check the computation from section 35.2 that planar geodesics are straight lines.

4. Consider the infinite cylinder of radius a with coordinates θ and z and the usual Riemannian metric given by

$$ds^2 = a^2\, d\theta^2 + dz^2.$$

a. Compute the Christoffel symbols.

b. Compute and solve the equations for geodesics.

5. Starting with the metric 35.1(2), show that on a sphere of radius a,

$$\Gamma^1_{22} = -\sin\varphi\cos\varphi, \quad \Gamma^2_{12} = \Gamma^2_{21} = \cot\varphi,$$

and that the other Christoffel symbols vanish.

6. Using Exercise 5 and Lemma 28.1, verify that great circles satisfy the geodesic equations 35.2(1).

Chapter 36

NonEuclidean Geometry

For millennia geometry was based on the axioms of Euclid. This chapter will take us on a voyage from the safe harbor of Euclidean geometry to the open seas of hyperbolic geometry, which violates Euclid's famous fifth axiom.

36.1. Euclid. About 300 BC, the celebrated Ancient Greek mathematician Euclid showed how plane geometry could be deduced from five postulates, such as "two points determine an (infinite) line." For hundreds of years, most mathematicians felt that the notorious fifth postulate was unnecessary, that it could be deduced from the others, but no one could do it. According to an equivalent version by Playfair in the Eighteenth Century, the fifth postulate says that

> *there is a unique line parallel to a given line through a given point not on the line.*

(Two lines are *parallel* if they do not intersect.) Here is Euclid's original formulation:

> και εαν εις δυο ευθειας ευθεια εμπιπτουσα τας εντος και επι τα αυτα μερη γωνιας δυο ορθων ελασσονας ποιη, εκβαλλομενας τας δυο ευθειας επ απειρον συμπιπτειν, εφ α μερη εισιν αι των δυο ορθων ελασσονες.

[That, if a straight line falling on two straight lines make the interior angles on the same side less than two right angles, the two straight lines, if produced indefinitely, meet on that side on which are the angles less than the two right angles.] Incidentally, notice how the Greek letter sigma at the end of a word is written ς instead of σ.

36.2. Spherical geometry. At first glance, spherical geometry seems to show by example that the fifth postulate cannot follow from the others. On the sphere, "lines" are great circles, and *no* lines are parallel, because all great circles intersect. However, spherical geometry does not satisfy the four, standard postulates either, because the geodesics are not infinite.

Incidentally, in contrast to Euclidean geometry, in which the sum of the angles of a triangle is π, in spherical geometry the sum of the angles of a triangle is strictly greater than π. For example, a triangle from the north pole to the equator, a quarter of the way around the equator, and back to the north pole has three right angles summing to $3\pi/2 > \pi$.

36.3. Hyperbolic geometry. We will now present the hyperbolic plane $\mathcal{H}$, a true example of how the fifth postulate can fail. It was discovered in the Nineteenth Century independently by Bolyai in Hungary, Gauss in Germany, and Lobachevsky in Russia. In 1823 Bolyai wrote to his father saying:

> *I have discovered things so wonderful that I was astounded...*
> *Out of nothing I have created a strange new universe.*

Define $\mathcal{H}$ as the upper halfplane $\{(x, y) \in \mathbb{R}^2 \colon y > 0\}$ with the metric:

$$ds^2 = y^{\ 2}\, dx^2 + y^{-2}\, dy^2. \tag{1}$$

You might worry that $\mathcal{H}$ seems to come to a sudden end at the x-axis, but as you approach the x-axis, distance blows up, so that this apparent boundary is actually infinitely far away. Indeed, the distance from $(0, 1)$ to $(0, 0)$ is given by

$$L = \int_0^1 y^{-1}\, dy = \infty.$$

The following theorem identifies the geodesics of $\mathcal{H}$.

36.4. Theorem. *Geodesics of $\mathcal{H}$ are semicircles centered on the x-axis, as well as vertical lines.*

Proof. Exercises 2 and 3 show that geodesics, parameterized by arclength t, satisfy

$$\begin{aligned} 0 &= \ddot{x} - 2y^{-1}\dot{x}\dot{y} \\ 0 &= \ddot{y} + y^{-1}\dot{x}^2 - y^{-1}\dot{y}^2. \end{aligned} \tag{1}$$

Put $p = dx/dy$; then $\dot{x} = p\dot{y}$, $\ddot{x} = \frac{dp}{dy}\dot{y}^2 + p\ddot{y}$. Then substitute from the second equation in the first to get

$$\frac{dp}{dy} = y^{-1}(p^3 + p).$$

Integrating by partial fractions yields

$$-\frac{1}{2}\log(p^2+1) + \log p = \log y + \log c,$$

$$\frac{dx}{dy} = p = \pm\frac{cy}{\sqrt{1-c^2y^2}}.$$

If $c = 0$, $dx/dy = 0$ and the curve is a vertical line. Otherwise let $a = 1/c$ to obtain:

$$\frac{dx}{dy} = \pm\frac{y}{\sqrt{a^2-y^2}},$$

$$x - b = \pm(a^2 - y^2)^{1/2},$$

$$(x-b)^2 + y^2 = a^2,$$

a circle centered on the x-axis.

Since the steps are reversible, there are no other geodesics. Alternatively, we already have a geodesic through every point in every direction. □

36.5. NonEuclidean geometry. As Figure 36.1 shows, in hyperbolic geometry (which does satisfy the first four postulates), the fifth postulate fails: through a point not on a given geodesic (semicircle) there are *infinitely many* geodesics parallel to the given geodesic. Mathematicians now appreciate planar, spherical, and hyperbolic geometry as the three great geometries.

Incidentally, in the plane, the sum of the angles of a triangle is π; on the sphere, the sum of the angles of a triangle is greater than π; and in the hyperbolic plane, the sum of the angles of a triangle is always less than π.

Exercises 36

1. Show by example that the sum of the angles of a spherical triangle can be arbitrarily close to π.

2. For the hyperbolic metric 36.3(1), compute that

$$\Gamma^1_{12} = \Gamma^1_{21} = -\Gamma^2_{11} = \Gamma^2_{22} = -y^{-1},$$

and that the rest of the Christoffel symbols vanish.

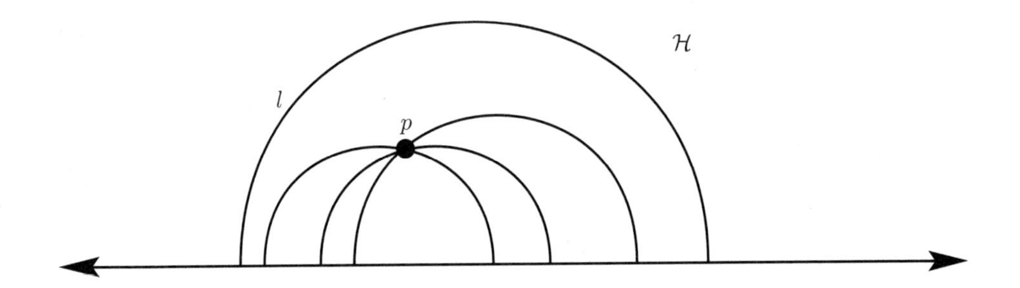

Figure 36.1. In the hyperbolic upper halfplane $\mathcal{H}$, where geodesics are semicircles centered on the x-axis, Euclid's fifth postulate fails: through a point p not on a given geodesic γ, there are *infinitely many* geodesics parallel to γ.

3. Use Exercise 2 to show that hyperbolic geodesics (parameterized by a multiple of arclength) satisfy 36.4(1).

4. Starting with 36.4(1), fill in any missing details in the proof of Theorem 36.4.

5. Give a direct proof of 36.4(1) as follows. Starting with the formula for length for any nonvertical curve,

$$L = \int y^{-1}(1 + y'^2)^{1/2}\, dx,$$

use 27.1(1) to obtain this first integral of the Euler equation:

$$y' = \pm\frac{\sqrt{a^2 - y^2}}{y}.$$

Now integrate to obtain circles centered on the x axis.

Chapter 37

General Relativity

During the late 1800's, a puzzling inconsistency in Mercury's orbit was observed.

Newton had brilliantly explained Kepler's elliptical planetary orbits by solar gravitational attraction and calculus. His successors used a method of perturbations to compute the deviations caused by the other planets. Their calculations predicted that the elliptical orbit shape should rotate or precess some fraction of a degree per century:

Planet	*Predicted Precession*
Saturn	$46'$/century
Jupiter	$432''$/century
Mercury	$532''$/century

Here $60'$ (60 minutes) equals one degree of arc, and $60''$ (60 seconds) equals one minute of arc.

Observation of Saturn and Jupiter confirmed the predictions. Mercury precessed $575''$/century. By 1900, the disagreement exceeded any conceivable experimental error. What was causing the additional $43''$ per century?

General relativity would provide the answer.

37.1. General relativity. The theory of general relativity has three elements. First, special relativity describes motion in free space without gravity. Second, the Principle of Equivalence extends the theory, at least in principle, to include gravity, roughly by equating gravity with acceleration. Third, Riemannian geometry provides a mathematical framework which makes calculations possible.

Figure 37.1. "Mercury's running slow." By J. Bredt, from Morgan's *Riemannian Geometry*, © Frank Morgan.

37.2. Special relativity. A single particle in free space follows a straight line at constant velocity, e.g., $x = at$, $y = bt$, $z = ct$ or

$$\frac{x}{a} = \frac{y}{b} = \frac{z}{c},$$

the formula for a straight line through the origin in 3-space. This path is also a straight line in 4-dimensional space-time:

$$\frac{x}{a} = \frac{y}{b} = \frac{z}{c} = \frac{t}{1},$$

i.e., a geodesic for the standard metric

$$ds^2 = dx^2 + dy^2 + dz^2 + dt^2. \tag{1}$$

Actually it is a geodesic for any metric of the form

$$ds^2 = a_1\,dx^2 + a_2\,dy^2 + a_3\,dz^2 + a_4\,dt^2. \tag{2}$$

Einstein based special relativity on two axioms:

Axiom I. The laws of physics look the same in all inertial frames of reference, i.e., to all observers moving with constant velocity relative to one another. (Of course in accelerating reference frames physics looks different. Cups of lemonade in accelerating cars suddenly fall over and tennis balls on the floors of rockets flatten like pancakes.)

Axiom II. The speed of a light beam is the same relative to any inertial frame, whether moving in the same or opposite direction. (Einstein apparently guessed this surprising fact without knowing the evidence provided by the famous Michelson–Morley experiment. It leads to the other curiosities, such as time's slowing down at high velocities.)

Einstein's axioms hold for motion along geodesics in space-time if one takes the special case of (2)

$$ds^2 = -dx^2 - dy^2 - dz^2 + c^2\,dt^2. \tag{3}$$

This is the famous Lorentz metric, with c the speed of light. We will choose units to make $c = 1$.

The Lorentz metric remains invariant under inertial changes of coordinates, but looks funny in accelerating coordinate systems.

For us a new feature of this metric is the presence of minus signs. It turns out that this is nothing to worry about.

This new distance s is often called "proper time" τ (Greek letter tau), since a motionless particle (x, y, z constant) has $ds^2 = dt^2$.

In Newtonian mechanics, a planet orbiting a sun stays in the plane determined by its initial position and velocity. This turns out to remain true in relativity, so we'll assume that $z = 0$ and focus on the x-y plane. If we also change (x, y) to polar coordinates (r, θ), the Lorentz metric becomes

$$d\tau^2 = -dr^2 - r^2\,d\theta^2 + dt^2. \tag{4}$$

37.3. The Principle of Equivalence. Special relativity handles motion—position, velocity, acceleration—in free space. The remaining question is how to handle gravity. Einstein's Principle of Equivalence asserts that infinitesimally the physical effects of gravity are indistinguishable from those of acceleration. If you feel pressed against the floor of a tiny elevator, you cannot tell whether it is because the elevator is resting on a massive planet or because the elevator is accelerating upward. Consequently, the effect of gravity is just like that of acceleration: it just makes the formula for ds^2 look funny. Computing motion in a gravitational field will reduce to computing geodesics in some strange metric.

37.4. The Schwarzschild metric. The appropriate metric for a planar solar system, assuming that the sun is a point mass in the center of otherwise

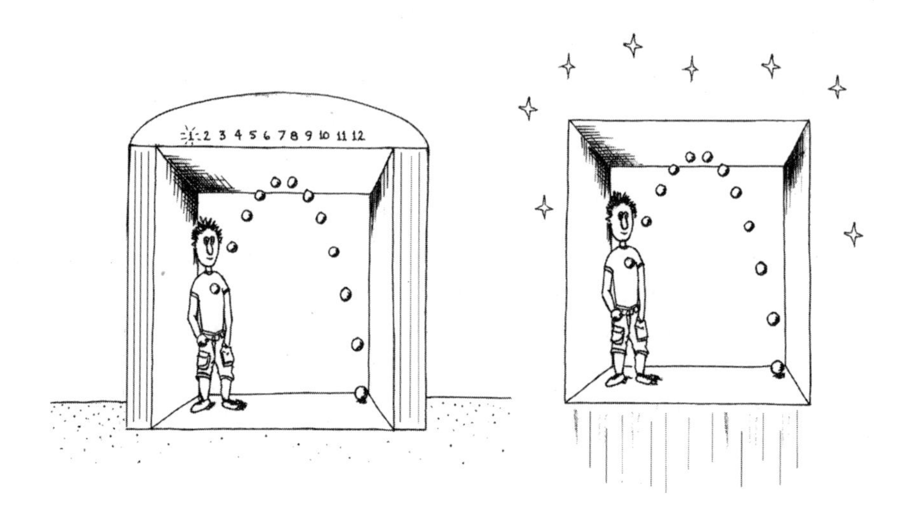

Figure 37.2. The Principle of Equivalence of gravity and acceleration: if you feel pressed against the floor of a tiny elevator, you cannot tell whether it is because the elevator is resting on a massive planet or because the elevator is accelerating upward.

empty space, turns out to be the planar "Schwarzschild Metric":

$$(1) \qquad d\tau^2 = -(1 - 2GMr^{-1})^{-1}\, dr^2 - r^2\, d\theta^2 + (1 - 2GMr^{-1})\, dt^2,$$

where M is the mass of the sun and G is the gravitational constant. Notice that if $M = 0$ (if the sun is removed), the Schwarzschild Metric (1) reduces to the Lorentz metric 37.2(4).

In Euclidean space, r represents both the radius and $1/2\pi$ times the circumference of a circle about the origin. Here it continues to represent the latter; hence the unchanged second, tangential term $-r^2\, d\theta^2$, exhibiting no distortion in the direction of the circumference. The distortions show up in dr and dt terms, in the radial and the time directions. In particular, time slows down in a gravitational field.

Notice that as r decreases to $2GM$, $d\tau^2$ blows up, creating a singularity past which nothing can escape: shrinking the sun to a point mass has created a black hole of "Schwarzschild radius" $r = 2GM$.

37.5. Mercury's orbit. Now we are ready to see what differences general relativity predicts for Mercury's orbit. The physics is embodied in the three equations for geodesics generalizing 35.2(1) in the Schwarzschild Metric 37.4(1). Three equations should let us solve for r, θ, and t as functions of τ. Actually, instead of the first equation for geodesics involving $d^2r/d\tau^2$, we will use the Equation 37.4(1):

$$(1)\quad -(1-2GMr^{-1})^{-1}\left(\frac{dr}{d\tau}\right)^2 - r^2\left(\frac{d\theta}{d\tau}\right)^2 + (1-2GMr^{-1})\left(\frac{dt}{d\tau}\right)^2 = 1.$$

To compute the two other geodesic equations, Exercise 1 computes from the Schwarzschild metric 37.4(1) that

$$(2)\qquad \begin{aligned} \Gamma^2_{12} &= \Gamma^2_{21} = r^{-1}, \\ \Gamma^3_{13} &= \Gamma^3_{31} = GM(r^2 - 2GMr)^{-1}; \end{aligned}$$

and the other relevant Christoffel symbols vanish. Hence (Exercise 2) the last two geodesic equations are

$$(3)\qquad \frac{d^2\theta}{d\tau^2} + 2r^{-1}\frac{dr}{d\tau}\frac{d\theta}{d\tau} = 0,$$

$$(4)\qquad \frac{d^2t}{d\tau^2} + \frac{2GM}{r^2 - 2GMr}\frac{dr}{d\tau}\frac{dt}{d\tau} = 0.$$

Integration of (3) and (4) (Exercise 3) yields

$$(5)\qquad r^2\frac{d\theta}{d\tau} = h \quad (h \text{ is some constant}),$$

$$(6)\qquad (1-2GMr^{-1})\frac{dt}{d\tau} = \beta \quad (\beta \text{ is some constant}).$$

Therefore (1) becomes

$$(7)\qquad -r^{-4}\left(\frac{dr}{d\theta}\right)^2 - r^{-2}(1-2GMr^{-1}) + \beta^2 h^{-2} = h^{-2}(1-2GMr^{-1}).$$

Putting $r = u^{-1}$ yields

$$(8)\qquad \left(\frac{du}{d\theta}\right)^2 = 2GM\left(u^3 - \frac{1}{2GM}u^2 + \beta_1 u + \beta_0\right),$$

for some constants β_0, β_1. The maximum and minimum values u_1, u_2 of u must be roots. Since the roots sum to $1/2GM$ (Exercise 5), the third root is $1/2GM - u_1 - u_2$, and hence

$$\begin{aligned} \left(\frac{du}{d\theta}\right)^2 &= 2GM(u-u_1)(u-u_2)\left(u - \frac{1}{2GM} + u_1 + u_2\right), \\ \frac{d\theta}{|du|} &= \frac{1}{\sqrt{(u_1-u)(u-u_2)}}[1 - 2GM(u+u_1+u_2)]^{-1/2} \\ &\approx \frac{1 + GM(u+u_1+u_2)}{\sqrt{(u_1-u)(u-u_2)}} \end{aligned}$$

because $GMu = GM/r$ is a small quantity ε and $[1-2\varepsilon]^{-1/2} \approx 1+\varepsilon$ (Exercise 6).

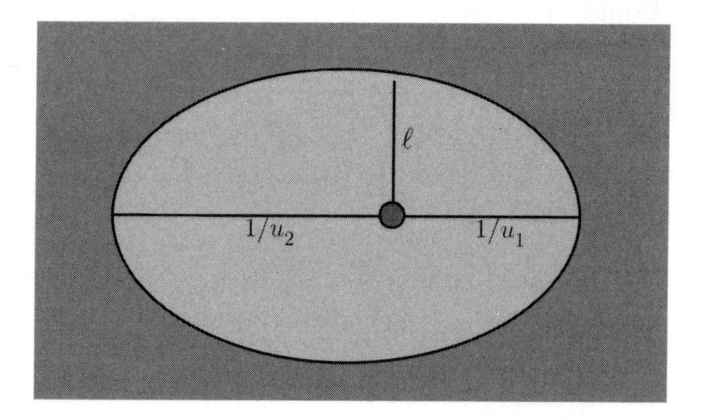

Figure 37.3. The classical ellipse

To first approximation, the orbit is the classical ellipse of Figure 37.3,

$$r = \ell/(1 + e\cos\theta)$$

(a standard formula of analytic geometry in most calculus books), or equivalently

$$u = \ell^{-1}(1 + e\cos\theta).$$

The largest and smallest values occur when θ is 0 or π:

$$u_1 = \ell^{-1}(1+e), \quad u_2 = \ell^{-1}(1-e).$$

The mean distance a satisfies

$$a = \frac{1}{2}\left(\frac{1}{u_1} + \frac{1}{u_2}\right) = \frac{\ell}{1-e^2}.$$

For one revolution (Exercise 7),

$$\begin{aligned}
(9) \qquad \Delta\theta &\approx \int_{\theta=0}^{2\pi} \frac{1 + GM\ell^{-1}(3 + e\cos\theta)}{\sqrt{\ell^{-1}e(1-\cos\theta)\ell^{-1}e(1+\cos\theta)}}\ell^{-1}e\,|\sin\theta|\,d\theta \\
&= \int_{\theta=0}^{2\pi} 1 + GM\ell^{-1}(3 + e\cos\theta)\,d\theta \\
&= 2\pi + 6\pi GM/\ell \\
&= 2\pi + 6\pi GM/a(1-e^2).
\end{aligned}$$

The ellipse has precessed $6\pi GM/a(1-e^2)$ radians. The rate of precession in terms of Mercury's period T is

$$\frac{6\pi GM}{a(1-e^2)T},$$

or, back in more standard units (in which the speed of light c is not 1),

$$\frac{6\pi GM}{c^2 a(1-e^2)T} \quad \text{radians.}$$

Now

$$G = \text{gravitational constant} = 6.67 \times 10^{-11}\ \text{m}^3/\text{kg sec}^2,$$

$$M = \text{mass of sun} = 1.99 \times 10^{30}\ \text{kg},$$

$$T = \text{period of Mercury} = 88.0\ \text{days},$$

$$a = \text{mean distance from Mercury to sun} = 5.768 \times 10^{10}\ \text{m},$$

$$e = \text{eccentricity of Mercury's orbit} = .206,$$

$$c = \text{speed of light} = 3.00 \times 10^{8}\ \text{m/sec},$$

$$\text{century} = 36525\ \text{days},$$

$$\text{radian} = 360/2\pi\ \text{degrees},$$

$$\text{degree} = 3600''.$$

Multiplying these fantastic numbers together (Exercise 8) we conclude that the rate of precession is about

$$(10) \qquad 43.1''/\text{century},$$

in perfect agreement with observation.

Exercises 37

1. Verify the Christoffel symbols of 37.5(2). (See 35.2(2).)

2. Verify the two geodesic equations 37.5(3, 4). (See 35.2(3).)

3. Check equations 37.5(5, 6) by differentiating.

4. Check equations 37.5(7, 8).

5. Check the algebra fact that if

$$u^3 - au^2 - bu - c = (u - u_1)(u - u_2)(u - u_3),$$

then $u_1 + u_2 + u_3 = a$.

6. Show that for ε small, $(1 - 2\varepsilon)^{-1/2} \approx 1 + \varepsilon$.

Hint: Let $f(x) = x^{-1/2}$ and use the fact that

$$f(1 + \Delta x) \approx f(1) + f'(1)\Delta x.$$

7. Check the computations of 37.5(9).

8. Check the final computation of 37.5(10).

9. What is the precession of the Earth's orbit due to general relativity? ($e \approx .0167$, $a \approx 1.5 \times 10^{11}\, m$.) Answer: $3.8''$/century. Why do you think that this result was not used to verify general relativity?

Partial Solutions to Exercises

Chapter 1.

1.1a. F, T, T, T.

1.2. Hint: It's easiest to answer, "For all x except..."

1.3. a, c.

1.9a. bijective.

Chapter 2.

2.1a. S is uncountable; otherwise $\mathbb{R}$, as the union of two countable sets, S and $\mathbb{Q}$, would be countable.

Chapter 3.

3.1. Converges to 0.

3.11. Given $\varepsilon > 0$, choose $N > 1/\varepsilon$. Then if $n > N$,

$$|a_n - 0| = |1/n - 0| = 1/n < 1/N < \varepsilon.$$

3.19a. $\mathbb{R}$.

3.21. False.

Chapter 4.

4.3. Let a_n be an increasing sequence in $\mathbb{R}$ which is bounded above. Since it is increasing, it is bounded below by a_1. By Theorem 4.3, it has a convergent subsequence. Since the original sequence is increasing, its terms lie between terms of the subsequence and therefore it also converges to the same limit as the subsequence.

4.6. Let a_n be a Cauchy sequence. By Exercise 3.18, a_n is bounded. By Theorem 4.3, a_n has a subsequence converging to a limit L. We claim that a_n converges to L. Given $\varepsilon > 0$, choose N so that for some $m > N$, $|a_m - L| < \varepsilon/2$ and for all $m, n > N$, $|a_n - a_m| < \varepsilon/2$. Then for all $n > N$,

$$|a_n - L| \leq |a_n - a_m| + |a_m - L| < \varepsilon/2 + \varepsilon/2 = \varepsilon.$$

4.7. 1, $+\infty$, $-\infty$, 1, 2, 1.

Chapter 5.

5.1. There are lots of examples, such as the greatest integer less than or equal to x; or the characteristic function $\chi_{\mathbb{Z}}$ of the integers. (Note that the exercise asks for a function defined on all of $\mathbb{R}$.)

5.7. Suppose that f and g are continuous. By definition, given a point p,

$$\lim_{x \to p} f(x) = f(p) \quad \text{and} \quad \lim_{x \to p} g(x) = g(p).$$

Hence by Proposition 5.6(2),

$$\lim_{x \to p} = (f + g)(x) = f(p) + g(p) = (f + g)(p).$$

Therefore $f + g$ is continuous.

5.12. The idea is that nearby integers have nearby values of f because there aren't any nearby integers!

Consider an integer p. To check the ε-δ definition, note that given any ε, you can take $\delta = 1$, and then $|x - p| < \delta$ implies that $x = p$, so that $|f(x) - f(p)| = 0 < \varepsilon$.

Using Proposition 5.7, note that any sequence of integers approaching p is eventually just $p, p, p, \ldots$. Hence the corresponding sequence of values is eventually just $f(p), f(p), f(p), \ldots$, which of course converges to $f(p)$.

Chapter 6.

6.1. $f \circ g(x)$ is x for part b, but not for part a.

Chapter 7.

7.1. Neither, closed, open, open.

7.5. $\{0, 1\}$, $(0, 1)$, $[0, 1]$.

7.10. Since "or" is always inclusive, it suffices to show that if p in S is not a boundary point, then p is an interior point. But if p in S is not a boundary point, then $p \in S - \partial S = \operatorname{int} S$.

7.11. First note that the interior of S is an open subset of S. Indeed if $p \in \operatorname{int} S$, then by definition $p \notin \partial S$, and hence some small ball B_1 about p is contained in S. A smaller ball B_2 about p is contained in $S - \partial S$, because if a boundary point of S were contained in B_2, a point of $S^{\complement}$ would lie in B_1. Hence by definition $\operatorname{int} S = S - \partial S$ is open.

Now let T be any open subset of S. T cannot contain any points of ∂S, because a ball about such a point contains points not in S and hence not in T. Thus, T is contained in $S - \partial S = \operatorname{int} S$. Therefore int S is the *largest* open set contained in S.

7.13. Suppose that $p \in \partial S$ is not an isolated point of S. Then a ball of radius $1/n$ about p contains another point a_n of S, and p is the limit of the sequence a_n. Therefore p is an accumulation point of S.

Chapter 8.

8.1. Take any sequence in the intersection $S \cap T$ of compact sets S and T. Since the sequence lies in S, some subsequence converges to a point S of S. Since that subsequence lies in T, some subsubsequence converges to a point t of T. Since any subsequence of a convergent sequence converges to the same limit, $s = t$, so that the limit point lies in $S \cap T$ as desired.

8.2. Let S and T be compact subsets of $\mathbb{R}^n$. Since S and T are both closed, $S \cap T$ is closed. Since S and T are both bounded, $S \cap T$ is bounded. Therefore $S \cap T$ is compact.

8.9. Immediately, from the definition, $\sup S \geq s$ for all s in S. If $a < \sup S$, then there is a point of $\bar{S}$ larger than a, and hence a point of S larger than a, so that a fails to satisfy $a \geq s$ for all s in S.

Chapter 9.

9.4. Since g is continuous, $g(K)$ is compact by Theorem 9.1. Likewise, since f is continuous, $f(g(K)) = (f \circ g)(K)$ is compact. Of course, you need to assume that $g(K)$ is contained in the domain of f.

Chapter 10.

10.3. Suppose that f and g are uniformly continuous. Since f is uniformly continuous, given $\varepsilon > 0$, we can choose $\delta_1 > 0$ such that

$$(*) \qquad |y_1 - x_1| < \delta_1 \Rightarrow |f(y_1) - f(x_1)| < \varepsilon.$$

Since g is uniformly continuous, we can now choose $\delta > 0$ such that whenever $|y - x| < \delta$, then $|g(y) - g(x)| < \delta_1$, which in turn implies that $|f(g(y)) - f(g(x))| < \varepsilon$, by $(*)$ with $y_1 = g(y)$ and $x_1 = g(x)$. Thus,

$$|y - x| < \delta \Rightarrow |(f \circ g)(y) - (f \circ g)(x)| < \varepsilon,$$

and $f \circ g$ is uniformly continuous.

Chapter 11.

11.1. The disjoint open sets $U_1 = (-\infty, 1/2)$ and $U_2 = (1/2, \infty)$ separate the integers into two nonempty pieces.

11.7. Suppose that f takes on two different values, y_1 and y_2. Choose an irrational number y_3 between y_1 and y_2. Then the open sets $U_1 = (-\infty, y_3)$ and $U_2 = (y_3, \infty)$ separate $f(\mathbb{R})$ into two nonempty pieces. This contradicts Theorem 11.4, which says that the continuous image of a connected set is connected.

11.8. By Proposition 11.2, Theorem 11.4, and Proposition 11.3, it must be an interval. By Theorems 9.1 and 8.2, it must be a closed interval.

11.11. First, suppose that S is totally disconnected. By definition, S has at least two points. If S contained an interval, a separation of two points of the interval would prove the interval disconnected, a contradiction. Conversely, suppose that S has two points but no interval. Let p_1, p_2 be distinct points of S. Since S does not contain an interval, there is a point p_3 between p_1 and p_2 not in S. Then the open sets $U_1 = (-\infty, p_3)$ and $U_2 = (p_3, \infty)$ provide the required separation to show that S is totally disconnected.

Chapter 12.

12.1. Let a be a point of the Cantor set. Let a_n be one of the endpoints of the interval of S_n containing a. (If one of the endpoints is a, choose the other one.) Then a_n is in C and a is the limit of the sequence a_n because the length of the intervals goes to 0.

Chapter 13.

13.3. Since $[a, b]$ is compact (Theorem 8.2), every continuous image is compact (Theorem 9.1), and hence every continuous function has a maximum (Corollary 9.2). This fact is used in the proof of Rolle's Theorem (13.3), to find a place where the derivative vanishes. Rolle's Theorem leads immediately to the Mean Value Theorem (13.4) and finally to Corollary 13.5, which says that on an interval where f' is always 0, f is constant. This final result will be a key ingredient in the proof of the Fundamental Theorem of Calculus (15.1).

Chapter 14.

14.1. [6, 10].

14.4. Let f be a nonnegative function with Riemann integral equal to A. Chop the interval up into identical subintervals with Δx small enough to guarantee that every Riemann sum is less than $A + 1$. Since every contribution is nonnegative, each subinterval contributes at most $A + 1$ to the Riemann sum. Therefore, on every subinterval f is bounded above by $(A + 1)/\Delta x$.

14.6. An unbounded interval cannot be chopped up into finitely many small intervals.

14.7. Yes. There will always be just one subinterval on which the chosen $f(x)$ could be 0 or 1, and as the subinterval width shrinks, the effect becomes negligible. Similarly, finitely many discontinuities are OK.

Chapter 15.

15.4. $-f(a)$.

Chapter 16.

16.3. A simple hypothesis is that f be bounded.

16.8. Given $\varepsilon > 0$, choose δ such that

$$|x - y| < \delta \Rightarrow |f(x) - f(y)| < \varepsilon/3.$$

Second, choose M such that $1/M < \delta$ and $C/M < \varepsilon/3$. Third, choose N such that

$$n > N \Rightarrow |f_n(k/M) - f(k/M)| < \varepsilon/3 \quad (k = 0, 1, 2, \ldots, M).$$

Now suppose that $n > N$. Given x, choose k such that $|x - k/M| < 1/M$, and, hence, by the Lipschitz condition

$$|f_n(x) - f_n(k/M)| < C/M < \varepsilon/3,$$

and, similarly, for f. Now

$$\begin{aligned}|f_n(x) - f(x)| &\le |f_n(x) - f_n(k/M)| + |f_n(k/M) - f(k/M)| + |f(k/M) - f(x)| \\ &< \varepsilon/3 + \varepsilon/3 + \varepsilon/3 = \varepsilon.\end{aligned}$$

Chapter 17.

17.4. Compute by switching the order of integration. Justify by Fubini's Theorem because the integral of the absolute value is at most the integral of $(\pi/3)(10)(1) = 10\pi/3$, which integral is $(10)(\pi/3)(10\pi/3) < \infty$.

17.8. In justifying the use of Leibniz's Rule, note that for y near 0,

$$\left|\frac{\partial}{\partial y}\frac{1}{x^2+y^2}\right| = \left|\frac{2y}{(x^2+y^2)^2}\right| \le \frac{1}{x^4} = g(x),$$

whose integral even from 1 to ∞ is finite.

Chapter 18.

18.1. Converges to 1/9.

Chapter 19.

19.1. 1, $-1/3$, $[-\infty, +\infty]$, $+\infty$.

Chapter 20.

20.1b. TRUE.

Chapter 21.

21.1a. Easiest to use ratio test.

21.5. $\sum_{n=1}^{\infty} = (-1)^{n+1}\frac{(x-1)^n}{n} = (x-1) - \frac{(x-1)^2}{2} + \frac{(x-1)^3}{3} - \ldots.$

Chapter 22.

22.1a. $\Gamma(5/2) = (3/2)\,\Gamma(3/2) = (3/2)(1/2)\,\Gamma(1/2) = (3/4)\sqrt{\pi}$.

22.2. $(8/15)\pi^2 r^5$, $(1/6)\pi^3 r^6$.

22.3. $4\pi r^2$, $2\pi^2 r^3$, $(8/3)\pi^2 r^4$.

Chapter 23.

23.2. $\sin x + 0 + 0 + \cdots$, as you would guess.

(If you graph the functions $\sin x$ and $\cos nx$ or $\sin nx$, you might see how things cancel to make the integral of the product vanish; or you could just believe the statement in 23.3 that "$\cos nx$ and $\sin nx$ ... turn out to be orthonormal functions," see also Section 23.7; or later you could just use Exercise 25.2.

23.3. By integration by parts,

$$\int_0^\pi x\cos(nx) = \frac{1}{n^2}[\cos(n\pi) - 1],$$

$$\int_0^\pi x\sin(nx) = -\frac{\pi}{n}\cos(n\pi),$$

$$\int_0^\pi x^2\cos(nx) = \frac{2\pi}{n^2}\cos(n\pi),$$

$$\int_0^\pi x^2\sin(nx) = -\frac{\pi^2}{n}\cos(n\pi) + \frac{2}{n^3}[\cos(n\pi) - 1].$$

Now use formulae 23.2(2).

Chapter 24.

24.1. $\frac{4}{\pi}\left(\frac{1}{1^2}\sin x - \frac{1}{3^2}\sin 3x + \frac{1}{5^2}\sin 5x - \cdots\right)$

24.2. $\varepsilon\frac{4}{\pi}\left(\frac{1}{1^2}\sin x\cos t - \frac{1}{3^2}\sin 3x\cos 3t + \frac{1}{5^2}\sin 5x\cos 5t - \cdots\right)$

24.5. First compute the Fourier series for $f(t)$:

$$\frac{1}{4} + \sum_{n=1}^{\infty} \frac{\sin \frac{n\pi}{2} \cos nt + \left(1 - \cos \frac{n\pi}{2}\right) \sin nt}{n\pi}.$$

Chapter 25.

No solutions.

Chapter 26.

26.1. $y = x$.

26.2. $y = x + 2$.

26.3. $y = C_1 x^4 + C_2$.

26.4. $y = -(2/3)x^3 + C_1 x + C_2$.

Chapter 27.

27.1. Plugging the cycloid solution (with $c = 1$, φ from 0 to $\pi/2$) into the formula for T yields $T = \Delta\varphi/\sqrt{g} = \pi/2\sqrt{g} \approx 1.57/\sqrt{g}$. For a straight line of length $L \approx 1.51$, there is constant acceleration $a = g/L$ obtained by multiplying g by the cosine $1/L$ of the angle. The velocity is at and hence $L = .5aT^2$. Therefore $T = L\sqrt{2}/\sqrt{g} \approx 1.63/\sqrt{g}$, about 4% longer.

Chapter 28.

28.1 It suffices to show that a small piece of geodesic is straight. Assuming that the geodesic is smooth, we may assume that a small piece is a graph $y = f(x)$. Its length is given by integrating $F(x, y) = \sqrt{(1 + y'^2)}$. Since F does not depend on x, the first integral F is constant, so that y' is constant, so that the geodesic is straight.

28.3. about 3317 miles, about 3741 miles (about 13% farther).

Chapter 29.

29.2. If you do this as a double integral with dx on the inside, this becomes just the standard 1-dimensional integration by parts for the inside integral.

29.6. Since $f\boldsymbol{V}$ vanishes on the boundary, by the Divergence Theorem

$$0 = \int \operatorname{div}(f\boldsymbol{V}) = \int (\nabla f) \bullet \boldsymbol{V} + \int f \operatorname{div} \boldsymbol{V}.$$

Chapter 30.

No solutions.

Chapter 31.

31.4. $0 = \delta A = \int (1/2)(1+(\nabla f)^2)^{-1/2} 2\nabla f \delta(\nabla f) = -\int \operatorname{div} \frac{\nabla f}{\sqrt{1+(\nabla f)^2}} \delta f$ by Exercise 29.4, so that $\operatorname{div} \frac{\nabla f}{\sqrt{1+(\nabla f)^2}} = 0$.

31.5.

$$\begin{aligned}
0 &= \operatorname{div} \frac{\nabla f}{\sqrt{1+(\nabla f)^2}} \\
&= \frac{\partial}{\partial x}[f_x(1+f_x^2+f_y^2)^{-1/2}] + \frac{\partial}{\partial y}[f_y(1+f_x^2+f_y^2)^{-1/2}] \\
&= f_{xx}(1+f_x^2+f_y^2)^{-1/2} - f_x(1+f_x^2+f_y^2)^{-3/2}(f_x f_{xx} + f_y f_{xy}) \\
&\qquad + f_{yy}(1+f_x^2+f_y^2)^{-1/2} - f_y(1+f_x^2+f_y^2)^{-3/2}(f_x f_{xy} + f_y f_{yy}). \\
0 &= (f_{xx} + f_{yy})(1+f_x^2+f_y^2) - f_x^2 f_{xx} - 2 f_x f_y f_{xy} - f_y^2 f_{yy} \\
&= (1+f_y^2) f_{xx} - 2 f_x f_y f_{xy} + (1+f_x^2) f_{yy}.
\end{aligned}$$

31.6. A surface of revolution has an equation of the form $r = g(z)$, where $r = \sqrt{x^2 + y^2}$. Differentiating $g^2 = x^2 + y^2$ implicitly yields

$$\begin{aligned} gg'z_x = x, \quad gg'z_y = y, \\ (g'^2 + gg'')z_x^2 + gg'z_{xx} = 1, \\ (g'^2 + gg'')z_y^2 + gg'z_{yy} = 1, \\ (g'^2 + gg'')z_x z_y + gg'z_{xy} = 0. \end{aligned}$$

Applying the minimal surface equation to $z(x, y)$ yields

$$\begin{aligned} 0 &= [(1 + z_y^2)z_{xx} + (1 + z_x^2)z_{yy} - 2z_x z_y z_{xy}]gg' \\ &= (1 + z_y^2)(1 - (g'^2 + gg'')z_x^2) + (1 + z_x^2)(1 - (g'^2 + gg'')z_y^2) \\ &\qquad + 2z_x z_y(g'^2 + gg'')z_x z_y \\ &= 2 + (z_x^2 + z_y^2)(1 - g'^2 - gg'') \\ &= (z_x^2 + z_y^2)(1 + g'^2 - gg''). \end{aligned}$$

Therefore, $gg'' = 1 + g'^2$. Substituting $p = g'$ yields $gp(dp/dg) = 1 + p^2$. Integration yields $p^2 = a^2 g^2 - 1$, i.e.,

$$\frac{dg}{\sqrt{a^2 g^2 - 1}} = \pm dz.$$

Integration yields $(1/a)\cosh^{-1} ag = \pm z + c$, i.e., $r = g(z) = (1/a)\cosh(\pm az + c) = (1/a)\cosh(az \mp c)$, which is congruent to $r = (1/a)\cosh az$.

Chapter 32.

32.4. Use $T = (1/2)m\dot{h}^2$, $V = mgh$.

Chapter 33.

33.2. Integrating the Euler equation conclusion $x'' = c_2/2c_1$ yields $x = (c_2/4c_1)t^2 + at + b$. Plug in the conditions $x(t_1) = 0$, $x'(t_1) = 0$, $x(T) = 0$ to get three equations in the three unknowns a, b, t_1. Use the second to eliminate a. Then subtract the first from the third to eliminate b. Now you can solve for $(T - t_1)^2$ to be $4c_1B/c_2$ and hence $t_1 = T - \sqrt{4c_1B/c_2}$.

Chapter 34.

34.1. Go back to (1) and solve for K from

$$K' - iK = -C_0 e^{it}.$$

Multiplying by the integrating factor e^{-it} yields

$$\begin{aligned} K'e^{-it} - iKe^{-it} &= -C_0, \\ (Ke^{-it})' &= -C_0, \\ Ke^{-it} &= -C_0 t + K(0). \end{aligned}$$

Since $K(T) = 0$,

$$\begin{aligned} 0 &= -C_0 T + K(0), \\ C_0 &= K(0)/T. \end{aligned}$$

Chapter 35.

35.2. $2\pi a \sin\varphi$, $2\pi a$.

35.4.a. all zero.

35.4.b. $\ddot{u} = \ddot{v} = 0$, $u = u_0 + at$, $v = v_0 + bt$.

Chapter 36.

36.1. A tiny triangle is nearly Euclidean.

Chapter 37.

No solutions.

Greek Letters

α	A	alpha	ν	N	nu
β	B	beta	ξ	Ξ	xi
γ	Γ	gamma	o	O	omicron
δ	Δ	delta	π	Π	pi
ε	E	epsilon	ρ	P	rho
ζ	Z	zeta	σ	Σ	sigma
η	H	eta	τ	T	tau
θ	Θ	theta	υ	Υ	upsilon
ι	I	iota	φ	Φ	phi
κ	K	kappa	χ	X	chi
λ	Λ	lambda	ψ	Ψ	psi
μ	M	mu	ω	Ω	omega

Index